I0393812

CONTENTS

INTRODUCTION---------------------------p.11
PREFACE NOTE ----------------------------p.25
DIMENS. BIOLOGIST'S TOOLKIT-----p.30

APPENDIX I. Animal Phenomenology-------p. 281
APPENDIX II. Dimensional Properties-------p. 282
APPENDIX III. Beautific View----------------- p. 283
APPENDIX IV. Typologies ---------------------p. 284
APPENDIX V. Organization Levels----------- p. 286
APPENDIX VI. Minerals-------------------------p.287
APPENDIX VII. Biologist's Gold--------------p. 288
APPENDIX VIII. E. Terrapin-------------------p. 289
APPENDIX IX. Consciousness------------------p. 290
APPENDIX X. Vocabulary----------------------p. 291
APPENDIX XI. Axioms--------------------------p. 292
APPENDIX XII. Princip-Puzzle----------------p. 293
APPENDIX XIII. Pre and Post-----------------p. 294
APPENDIX XIV. Critique----------------------p. 295

BIBLIOGRAPHY --------------------------p. 298

INDEX ------------------p. 303

BIO -----------------------P. 326

Nathan Coppedge

A NOTE:

This text is designed neither specifically for philosophers, nor specifically for biologists. It apprehends that there may be some open categories in the study of biology, which determine that the study is as open-ended as nature. Although professional biologists, geneticists, and chemists are busy with a broad range of technicalities, this text anticipates what may be a sadly overlooked opportunity: to make biology scientific for logic and the liberal arts. I hope my readers will appreciate the considerable efforts and due attention that went into the so-called 'un-investigated corners' within this text. I consider the achievement a notable exception from the norm.

HAVE A NEW DISCOVERY THAT CONTRIBUTES TO DIMENSIONAL BIOLOGY? / PARTICULARLY USING CATEGORICAL STRUCTURES? SEND AN E-MAIL TO: CONTACT [AT] NATHANCOPPEDGE.COM

2

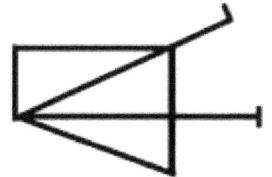

THE DIMENSIONAL
ENCYCLOPEDIA
VOL. III

Nathan Coppedge

Nathan Coppedge

DIMEN-SIONAL BIOLOGIST'S TOOLKIT

Or, Coherent Theories of Macrobiology

NATHAN COPPEDGE

Nathan Coppedge

DEDICATION

This book is devoted humbly to life's ultimate exceptions, without which we very likely would fail to have an idea of ourselves, and without which we may perhaps lack a future.

Nathan Coppedge

Introduction

I.

This is a book of ideas, not facts. So I wish to situate this text more precisely in the terrain of theories than in traditional territory of knowledge: what may be after all a highly contingent context of life forms and ecosystems. Dimensional Biology---what I call dimensional biology, or the biology of dimensionism---is a discipline founded on the idea that biology can be abstractive, that general theories can be formed about biology.

So it was with a little hesitation that I admitted that biology is the appropriate subject of the third volume of my encyclopedia. Although biology is widely studied by students at a basic level, few have the grasp of the real workings of biology in the sense that it would relate to complete systems. Indeed, according to the evidence, systems are highly dynamic, and are themselves highly contingent: evidence of this includes knowledge of pathogens, the existence or non-existence of certain forms of microscopic life, and the potential reality or non-reality (or relative degree) of speciation in the local universe.

Dimensional biology is then predicated on not only the notion that science has made significant developments, but that these categories upon which the context con- tinges, pose real viabilities in the search for an abstract formulation of the nature of life. In essence, the book is an ambitious one.

A short list of things needed to be accom- plished in this book is a prospectus of forms of life which have not yet been dis- covered, general patterns of adaptability, and general descriptions of how biology functions across plants and animals. This is no small task, by any means.

For one thing, predictions about the na- ture of alien life have varied widely from the bizarre to the simplistic. And while these two aspects seem like tentative keys to the process of universal biology, making hazardous guesses is a difficult dilemma, proven by the recent scientific vacillation first about whether dinosaurs have bright colors, and secondly the surprising possi- bility that they might have feathers instead of scales. So by no means am I approach- ing this subject from an angle of assumed predictability.

I am using the concept of 'second nature' as a clue to the concept of the categoriza- tion of organisms, because it allows a plat- form between one category and the next, say, a maximal shared number of traits of fifty percent between major categories. This has the advantage of potentially---in light of certain categorical techniques---revealing new forms of life of which human science as we know it today was not aware.

Although the most surprising thing about this text may be its look at speciation, I have taken some pains to look creatively at the function of a wide variety of organisms.

I have introduced a number of major theoretical premises. One of them is the notion of *vast exceptionism*, summed up in a relation between dimensional complexity, a corollary of adaptive complexity, and a general thesis of exceptional adaptation. In other words, [1] Although I do not take a religious viewpoint, I see that the acceptance of radical simplicity would present itself as a diabolical fact which denies the sheer potential of nature and civilization. [2] If nature is physically complex, it also follows that adaptation has a corresponding complexity. Organisms adapt exponentially for every available complexity in nature. [3] Exceptional adaptation thus takes place first as a function of the laws of nature, and secondly as a function of the complexity of the organism. In what I call dimensional biology it is thus worth interpreting the significance of variables of functionalism which are almost purely abstract --- because they have bearing on in-

13

terpretations of theory. I am arguing against what I call the 'meme-like functionalism' in traditional Darwinism and social behaviorism. Big labels like evolution are less emphasized, and may even no longer be necessary to grasp nature ---or civilization --- coherently.

Now it would behoove us to take this assessment a step further, and make further premises:

[4] A second iteration of the corollary is to make statements about what it means to reach advanced adaptations. Biological exceptions of this kind present themselves as separate facts--- differentiated from laws, or perhaps even equivalent to them. Such could be said of the so-called social contract of J.J. Rousseau fame, or Moore's Law in computing. As civilization progresses in its utilization of resources, it also increases its ability to create laws of nature, *Logics* with *no inherent impact on nature*.

In this sense of technology or bio-organization, nature itself has adapted, if only by the principle of paradigmatics. In such a system, one can't afford to not be a realist. Realism involves radical acceptance of new paradigms, offset by knowledge of survival. In some sense every advantage is a way of synthetically adjusting the properties of nature, or if not, then hierarchically modifying the sense of individual, group, or social significance.

[5] Synthesizing the Future: In the context of the most synthetic organisms (such as cyborgs or A.I.), functionality is also by that *definition* synthetic, suggesting the very far end of the current paradigmatic shift: an age in which *perhaps a high degree of cognitive function will replace mutation*. This is the onset of a new age in which genetic, physical, and cognitive modifications are commonplace. The thing to predict is whether paradigms will in fact be paradigmatic---or instead, only literal.

II.

There is a need for a degree of qualification in what is meant by exceptional biology, and whether it exists merely as a 'thought experiment' with no moral function, or whether some group or social concept of psychology is adapted to account for the qualification; A rather famous article titled "Epistemological issues in omics and high-dimensional biology: give the people what they want", illustrates the conceptual hurdles in adopting even the most advanced conventional approach to biology:

> "For ease of illustration, suppose there are two treatments, treatment A and treatment B...[the differential equation] is evaluated using a test statistic computed on observed data and a P

value computed under some assumed reference distribution (i.e., a distribution of a test statistic when [that] is true...) Valid P values will be obtained if the required assumptions for the test are met, and they may not be valid if one or more assumptions are not met. In microarray experiments, sample sizes are often not large, and the expression levels may not be normally distributed. Alternative methods for obtaining a reference distribution fall under a category of methods sometimes called non-parametric or distribution-free. These terms are slightly misleading in that parameters may still be estimated using these [earlier] methods, and some assumptions regarding the distribution of data are required. The hypothesis-testing framework above refers to a classical frequentist approach. There are alternative approaches such as Bayesian...[Another technique] has the disadvantage of being computationally intensive. There are marked differences in how such approaches are implemented, and some confusion and uncertainty remain. [Scientists who use these methods] often do not discuss these

*issues nor state why one [such]
approach ... is chosen over another"*

-----(Tapan S. Mehta, Stanislay O.
Zakharkin, Gary L Gadbury, and David B.
Allison; Physiological Genomics, January
2007, Vol. 28, No. 1; p. 24-32)

Readers of David K. Lewis have sometimes
posed that science is limited where it depends on *alethic abilities*---a vocabulary
which refers to the use of specialized tools
above the science of understanding both
the context and the significance of technicalism. It might be convenient to have
some secondary sets of categories which
clarify the specifically *meaningful* opera-
tion of biology, either technically or emo-
tionally. It should not be assumed that
meaning is purely mathematical (certainly
not statistical), raising questions about the
nature of *use*; It may well suggest that information may be designated by many examples which are not present in the most
technical concepts of nature as it has been
portrayed. It may be useful to project that
there is a recurrence of previous themes
from nature, however subtle or secondary
this relevance might be, that provides a
framework for considering that information is far more broad in its ultimate scope.

While it depends on the specific, these examples may be more general, both more
abstract and more tangible, than the ex-

perimental sciences have been able to pro-
tract. We should not assume that because
the most general notions conceivable are
irrelevant to science, that every generality
in-between must lack relevance.

So, too, we should not assume that every
scientific method is meaningful, only because it has committed to a material form
of any degree of technicalism. *Meaning, it
seems, is another dimension of analysis.*
And, this is a major claim: *it is not automatically implicated within scientific information.* And while it may be formal, it
may also have its own formality, and
where it concerns information that is not
divulged by conventional methods of any
degree of advancement, then so too it offers its own method of analysis.

However, my bold perspective depends on
the utility of exclusiveness, and so it also
depends on the existence of opposites in
nature. By opposites what I mean are elements. Opposites exist in at least two
types: axiallary opposites are pure opposites which contradict one another. Contingent opposites on the other hand, provide territories for each-other's existence.

The theory does not predict that oppositeness necessarily creates conflict: indeed,
comparison might be a more accurate
word. Fortunately, ecological diversity has
become a sacred principle of biology and
horticulture. So there is a degree of justice

in conceiving that once an exclusive set consists of real abstractions of nature, that these abstractions are ecologically compatible, because all organisms are adapted with their own survival in mind. The method of deduction does not posit real conflict as much as an observation of real patterns in nature, patterns which must exist on the basis of differences, or else, so to speak, 'all would cohere'.

I will push away the pessimists from my dimensional science, because they do not accept the future of biotechnology. Nor do they accept the fundamental principle of biological diversity. I am not a pessimist, but I am a kind of fatalist about some things which we cannot easily control. Biology in my mind concerns this unusual condition of consciousness---simultaneous compromise and realization, which has already been familiar to the *biosophists*.

It is not just the organisms of nature which may inhabit categories, it is also the modalities and qualities of individuals and societies. I will shy away from the formulation in which species embody already the complexity which they fundamentally dominate. Instead, it should be ruled that much of the descriptive potential of an organism exists as information, and not biology. Individual organisms often reach toward a sense of self-identity that is not completely fulfilled, which seems to prove that information is dominant.

III.

Species can be studied in terms of their exceptional individuality, or they may be studied as derivatives of one or several binding equations. One aspect is to find the quantitative or dithered (multiplex or statistical) proof for general trends in nature. This may be understood with or without referring to actual quantitative data. Indeed, the size of populations is ubiquitous for all successful groups, except under very binding circumstances. The shyness to treat animals as legalities limits the potential of the view that some genuine species lack populations. Further categories may be found by defining factors such as reproduction, competition, or symbiosis which relate to the constancy of populations.

A second type of approach which I will call gestalt biology, is to study individual organisms as systems which have a metaphorical or symbolic role as representatives of generalized cross-species concepts. In this case, the realization of success is not by representation in a simple sense, but rather *representation in a qualified sense*. Modalities of individuals are used as benchmarks for the functionality of society as a whole, creating a kind of *ideational* biology. The study of individuals also offers the prospect of understanding exceptional characteristics of biology. The trend

to generalize the individual, however, must
be counter-acted by the principle that,
given a certain amount of prior development, activity is always a form of functionality, however primitive. Rejecting this minor thesis returns the equation to a psychological rather than biological standpoint. Typologies can be uncovered which
relate to the quantification of properties of
individuals, quantification of social characteristics, or quantification of exceptional
survival mechanisms, whether they are social or individual.

A third type of approach is not quantified-
descriptive or local-modal, but instead describes the relative function of society in
terms of available ideas. In this case, variables involve functional paradigms and
compatibility with those paradigms. If
there is an alien species that can survive
without war or labor, then one asks the
question as to whether one can imitate the
species, or become symbiotic. This hooks
into not only basic survival question, but
also individual perspective and existential
survival. The role of leaders becomes
deeply symbolic of the function of biology,
unless government is seen as taking place
by piecemeal arbitration by individuals. In
this view there is an appeal of the relative
as well as the conceptual view, just as in
Volume I.

This third category emphasizes the value
of general ideas and transcendental aware-

ness as a function with production, language, communication, and other forms of power (defined specifically). The second category emphasized the exponential value of specific properties, which I suggest is a kind of specific-specific relationship, while the first category emphasized that generally implemented properties have a general effect, and other similar ideas. By implication, a fourth category relates to the role of specifics in relation to generalities, such as consciousness and spontaneous evolution.

SUMMARY OF THE INTRODUCTION

Summarizing the four categories yields a context which is both complex and easily described:

[A] Social Cohorts, as a Function of Qualities
[B] Species, as a Function of Materials
[C] Nature, as a Function of Modes
[D] Development, as a Function of Incident

This quadra, actually consisting of four quasi-independent theses, provides a foundation for an adequately complex, and therefore reducible formula for nature. By reducing general dynamics, specific modules can be reached. A modular quality of this science both sets it apart from advanced conventional techniques, and defines that, outside of abstraction, its primary function is as a secondary application. However, it is my hope that concepts introduced within this book will prove useful in spite of this.

Now, an initial word before the main text.

Nathan Coppedge

PREFACE NOTE

A turning point.

It would do to describe how my general theory of biology is based on a theory of happiness. For it appears that the human motivation for knowledge depends fundamentally on human happiness. Even attempts to correct human error study the presence or absence of happiness. And certainly, without human happiness there would be little effort to correct nature, except by promoting a happiness which had been earlier realized.

A second type of theory of biology takes happiness as a mode of functionality, and will develop interpretations of more physical theories. Such is the case with bees, which are now understood to be an ancient image of what humans call 'business'. While this may seem like a simple perspective, the existence of an allegory for business, or even the business of happiness, is a highly useful claim. Bees become one image, the image of mechanical biology.

So, as a background, there is a tableau of human happiness and the meaningful activity of animals, an image which I warrant is not always accurate to human psychology, but which offers the prospect of biological futurism. Another aspect is the mechanical nature, which, even if borrowed from bees, has implications to a broad range of biological categories.

Nathan Coppedge

"I do not declare great things about the order of nature: I only ascribe writing on the wall"

   ---The Hesitant Scribe

"Without his natural instincts, the common creature [including the writer] would be defenseless"

   ---Often Attributed to Darwin

THE

DIMENSIONAL

BIOLOGIST'S

TOOLKIT

Nathan Coppedge

============[**A**]============

Adaptation—Adaptation in humans begins with four successive levels, if we ignore reproduction: (1) Comfort. Like a baby deer, the human seeks comfort from the environment, which determines its will to survive. (2) Adaptation to Security. Grown women often seek marriage for security, while grown men seek jobs, arguments, and weapons. (3) Adaptation to Sickness. Being immune to sickness, someone can live a long life and be productive. Wellness divides functional people from dysfunctional people. (4) Adaptation to Radiation. Even functional people are not always adapted to radiation. Adapting cures cancer (a disease that strikes late in life). Adapting to radiation requires selecting body modifications and actively controlling their progress, leading to full adaptation to the environment in evolutionary terms.

Adaptivity Patterns -

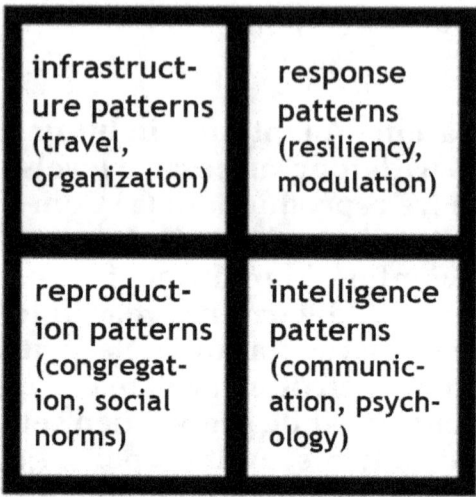

infrastruct-ure patterns (travel, organization)	response patterns (resiliency, modulation)
reproduct-ion patterns (congregat-ion, social norms)	intelligence patterns (communic-ation, psych-ology)

(Read All Quadra Diagrams Counter-clockwise from Upper Right).

Below: *Advanced Adaptation:*

New Advanced Function	Adaptive Response
New Related Functions	New Organz. Paradigm

Adding Concepts to Biology - Today an added concept must go beyond interior, exterior, ulterior, or extensory. It might, for example, explain communicability between differing surfaces. It might find a new purpose for multiple different combined units. It might find a new central-nervous function. Even in these areas, the evidence provided must be somewhat conclusive to merit scholarship. The general importance of a new system, according to analysis, has to do with the following: (1) Inter-surface concept, whether they are within or outside the body, (2) New information functions, and: (3) New coherent functions. Further areas can be extrapolated from these, such as initially: (1) New organic functions, (2) Information events, or (3) New paradigm concepts. However, these secondary cases are often considered too extreme, even if the first set was not. A more conservative set would look like the following: (1) New chemical information, such as DNA functioning, (2) Specific correspondence between DNA and behavior, such as medical diagnoses, and (3) Understanding of brain chemistry that reveals a range of functions. It is important to understand how conservative this final list is in comparison to the first or second. Discoveries about the physics and 'concept' of the system really are often more advanced than any applied chemical model, and may be even easier to reach.

Advanced Brains -

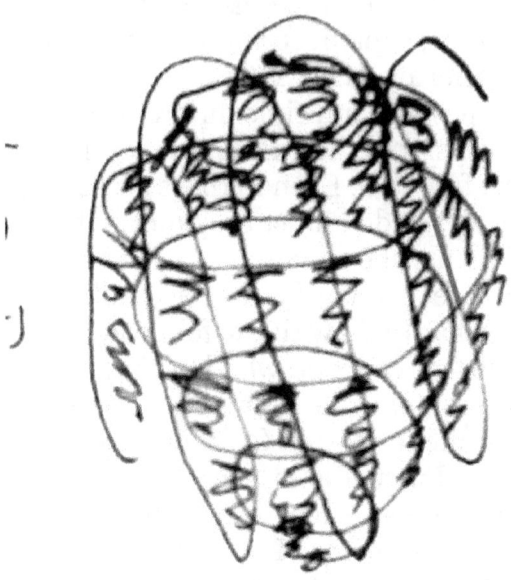

Some have thought that it is easy to imagine a more advanced brain structure, but that rarely would it achieve the same level of efficiency as the human brain. The major detractor to this theory is the concept that natural selection has thrown a wrench in the works, de-valuing intellectual enjoyment, and over-valuing senseless violence. Some would say we are beyond that age, but if the violence holds precedent in biology, it may be difficult to produce things like genuine patient wisdom, lucid context-independent intelligence, or unrepentant virtue, in the current brain structure, and under the view that reaction time and other features resulted from sce-

34

narios of stark survival. Then again, "When in doubt, overclock the system" so the saying goes. Many features mostly absent from the current human brain could be assimilated in the case of a much higher intelligence, an intelligence that takes various acts of pure genius as ordinary activities. The above diagram shows my attempt to design a brain for advanced intelligence. It has more complexity, more regularity, and a mechanism has been developed to organize by spiral instead of dividing into two lobes. Thus, the primary regulation occurs through reference between any individual 'zion' or energy/pleasure interpolator, and the overall reference structure represented by the entire group of zions. This occurs through the logichord, a spiral loop of communication matter which serves functions like organizing and enhancing information. In addition, there might be a central regulatory organ with the same importance as the heart or brain in a human, perhaps serving a psychic or emotional purpose. A similar function is already played in the human brain by the hippocampus, although perhaps in a more advanced brain this function would also be correspondingly advanced.

Advertising Skin Deep --- Beginning with scars and ear, lip, and nose piercings, the first thesis of intentional self-harm should be that it advertises material strength. But one should not assume that

that is the single and exclusive function of the injuries. Indeed, even clothing has in recent years been associated with perforation, malleability, and what some would call the '*ephemere.*' Some would say that the function is largely unconscious and actually exhibits social weaknesses. Others would say that torn clothing has the very same value as a battle scar, and even an intellectual cachet as a statement of irony and vulnerability associated with emotional development. In any case, the presence of such a cohort exhibits obvious patterns of coincidence with the military strength of a society, and the development of sufficient textile industry to support exhibitionism. Otherwise, the association is innately primitive, carrying with it the assumption of an automatic understanding, such as fitness and survival.

Alien / Xenoid Gestation - It may be useful to reduce human gestation to four critical somewhat early stages, many of which occur while the baby is still being nurtured by a mother. These stages ideally are also held in common with most advanced alien species, at least in their earliest stages. Thus, it becomes necessary to separate the stages from factors such as warmth or cold, while still factoring in those things which make a given gender and species unique reproductively. To do this, I have done a little research into the perspectives of human organisms, with the principle that early development is future-oriented in a quasi-Platonic sense. However, the goal as I said was to reduce gestation to four categories for every gender, and to provide a method that would work for any given species, even off of planet earth.

Here is my result for humans:

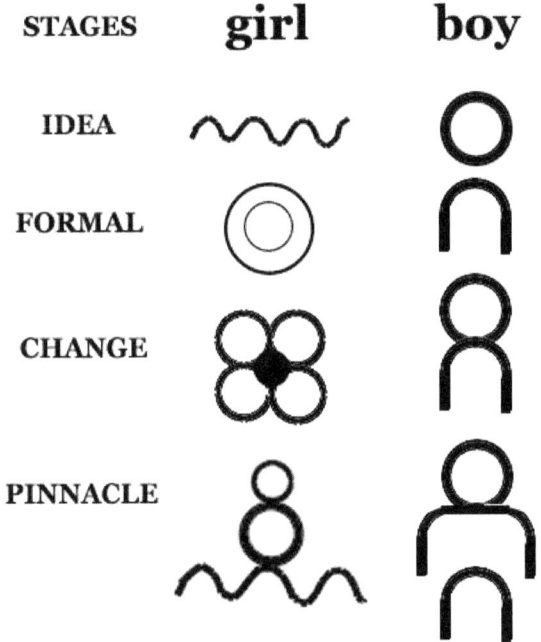

STAGES	girl	boy
IDEA		
FORMAL		
CHANGE		
PINNACLE		

In early life the girl sees the cosmic web of life. The boy sees an egg, which is an idea of the ultimate for him. The girl then focuses on the becomingness of life, as exemplified by the complex egg. The boy finds his legs. Then the girl, during pu-

37

berty, feels fat and connected with the world. The boy feels a little bit bigger. The girl, when she grows up, may feel pregnant, and once again feels connected with the web of life, although this time it appears in the form of the society which she knows. The boy, if he is lucky, feels like a man.

Here are my results for two hypothetical extraterrestrial species:

STAGES	BALZOG	BREZNIN
IDEA		
FORMAL		
CHANGE		
PINNACLE		

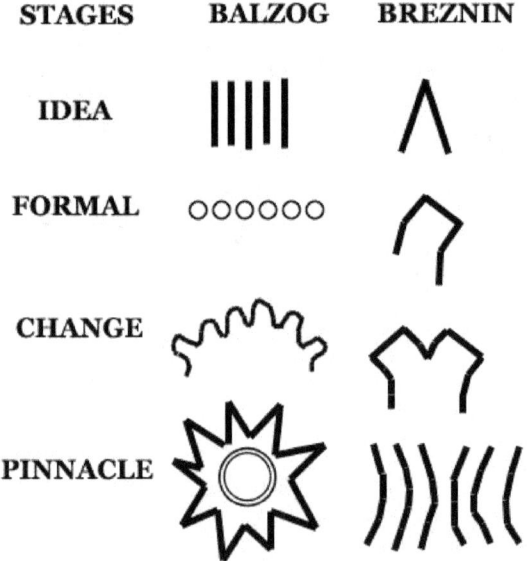

I am imagining a largely flaming planet that has developed two organisms adapted to reproducing here. The Breznin are emotional, metallic, sculptural looking entities.

The Balzog are seasonal, caustic, fertile entities. In the second stage the breznin begin to curve and emit chemicals in response to the downpour of the Balzog. The Balzog become fertile from the nutrients from the Breznin, and change into tiny seedpods. The seedpods cluster at the base of the Breznin, and begin to root themselves in the flaming turf. The Breznin undergo a molting process, expending the used parts, and gradually appearing as very peely metal plates which form massive space-occupying structures. Finally, the Balzog envelop the Breznin, forming a cluster of used seedpods at the base of the peely structure, which somehow allows the Balzog to gestate into downpours.

Alien / Xenoid Phenotyping - There is a tendency through recorded history to forget the exact nature that formerly identified a species or anthropological family to humans. It is my assessment that these were originally highly selective traits: individual characteristics like a high or low brow, much hair, or large feet. Inevitably as species begin to be considered as larger and larger aggregates, those traits begin to favor the generic rather than the particular. Because of intensive metaphysical study, we can identify characteristics which are common of most if not all organisms of a phylogenic lineage. The thing that binds one organism to its environment is what I take as my example of the myth-factoring or phenotype assessment of any organism. I found the following for plants and animals on earth, which I take to be Xenoid Type 0:

Plants: Hair-based.
Animals: Foot-based.

Phylogenic choices from here indicate (1) Different roles are possible, and (2) Other characteristics are possible. This leads to the broadest possible range of alien / xenoid alternate profiles, because, just as Immanuel Kant says, we get to the root of the matter.

Xenoid Profile 1: Animals are all hair-based, and plants are foot-based (seems improbable).

Xenoid Profile 2: We reject the animal and plant categories in favor of greater ambiguity. Animals are plants, and plants are animals. This could lead to plant brains and efficient animals.

Xenoid Profile 3: New definitive categories exist, with similar physics but different roles. Descriptions like 'feet are like [all] plants' or '[all] hair-species are like plants' become necessary. All of some major species resemble a specific type with much variation.

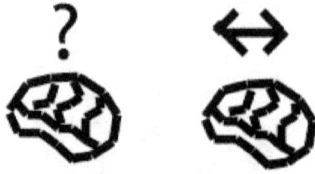

Xenoid Profile 4: An alternate landscape creates different laws of physics. These organisms break rules we're accustomed to following. 'Foot' and 'brain' lose all meaning [In the drawing, a brain is used for locomotion instead of thinking]. Where organisms appear to resemble ours, the comparison loses all meaning. Identical appearances are used for separate functions. The greatest similarity is the capacity for variety. Organisms are joined only by rare glimpses of common experiences, such as having stored energy or making use of space.

Xenoid profile 5: Instead of having slightly different laws, these organisms have none of the same laws. They can be theorized or simulated, but even believing in them requires a stretch of the imagination. These seem less probable than ghosts or boogie-

men, but it may just be a physical difference, such as having a different type of
space-time. Some of these organisms may
even look like us. But if the laws are not
analogous, the possibility grows unlikely.
These organisms may resemble the organisms we're used to (including alien species
and jellyfish), only with modification for
different laws. They may be much larger or
much smaller, for instance. Or they may be
like humans that can survive an atomic
blast. It is hard to predict what characteristics they might have. Metaphysical differences emerge like 'virtual' flowers,
'powerful' ants, 'space' trees, etc. if these
types have not already emerged, they
would emerge here, through a metaphysical or space-time / latent energy / conditional energy difference.

Amphibians - Amphibians are a case in
point --- perhaps the most frequently cited
example----in which categorical evolution
took place. One perspective is that evolution of this type is a rarity, and consequently it hardly represents a universal
case. That is, it does not respond to all the
factors necessary to become human or post
-human. Nonetheless, it is clearly an important stage for our species. Another perspective attempts to apply the metaphor of
amphibianism. Perhaps there are infinite-
-or some large number---of such stages
(like amphibianism!) all of which lead to
high levels of evolution. According to this

dramatic view, it is only a matter of how much you are allowed to stretch the metaphor. Surely the human mind can conceive of at least a step beyond the human----if amphibians were imaginative enough to step onto land! Clearly this second type of thinking leads to a dimensional view. Contexts such as virtual reality versus reality, and third or fourth dimension versus higher dimensions come to mind. These kinds of comparisons are what must be searched out if there is to be a new form of evolution----unless the evolution, as one trope has it, just involves something new and original, like 'how to be a robot' etc.

Animal Intuition -

The functioning of the animal brain has been widely cited in psychology and neuroscience. Often animals have adapted to have quicker, more instinctive responses to the most usual occurrences, and may react with shock if certain patterns are not followed. The prevalence of shock is said to diminish with the incidence of the primate brain, although shock is still triggered sometimes by social events such as death or mating conflict.

The subject of animal intuition is inevitably associated with the animal brain, and I will not introduce any supernatural theories on the matter. What is interesting to me is the psychological intuition, which is the role that the animal brain plays as a part of the human brain. I am interested in why someone looking for the word patois will often find the words 'tincture' or 'tapioca'. What holds the person (or for that matter, chimpanzee) back, from achieving a word selection? It might easily be a matter of mating selection, quality of life, or forms of stimulation.

What is the role of the choice to find the right word, is another question. In some cases, people consider finding a word to be utterly incidental, whereas in other cases, such as a spelling bee, people consider word-choice to be a form of selective behavior. One theory is that word-choice is directly analogous to a functional brain. In this view, all intelligent people (except those who cannot speak), display vocabularies which are established at their own level of intelligence. However, there are a number of obvious exceptions to this. Someone who slurs in their speech may have difficulty speaking, but it is possible to imagine that the same person could also have a strong vocabulary when writing. Also, someone studying physics may have a highly technical vocabulary which is hard to integrate with say, an English major's vocabulary, and yet the physics major may be a genius.

A second theory is not that words connote intelligence, but that they connote one's level of being socially accepted. Under this view, it is possible to be a highly intelligent person who is not integrated, because one has not been socially accepted. The obvious corollary is that animal intuition concerns things such as how vocabulary, or in the case of birds, singing or nest-building, or whatever else, may be understood, and thus adapted to.

There is a third theory, which is that intelligence including intuition is an arbitrary, acquired trait that has little more fated importance than a Barbie doll or the cracker jack that happens to fall on the wood flooring. Intuition may concern arbitrary things, or it may concern social things, or it may concern intelligent things. And there may be still other theories.

Animals [Profile]-

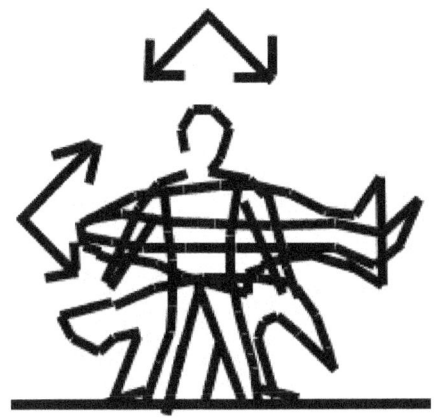

Every animal has four functions:

[1] ACCESS FUNCTION
Meeting species-level requirements.

[2] IMITATION FUNCTION
Such as occultation:
Defense from the sun,
and camouflage.

[3] COMMITMENT FUNCTION
Such as display or hiding.

[4] SURVIVAL FUNCTION
Such as dedication or persuasion
(and war and peace).

Antecommunication - An attitude for communication with foreign intelligences. Essentially one must 'ante' or offer the concept of communicating, independent of its definition. Conceptually, in the broadness of scope, this also means accepting the prospect that non-communication is also a form of communication (yet not by narrowing the definition of the meta concept). Thus, what follows is a mode in which complexity involves communication, but there are many simple platforms for doing so. A technical way to phrase this is that communication is arbitrary only by terms of negation. Furthermore, it must be accepted that multiple categories of non-communication exist, with influence upon the complexity of any insensitive approaches.

Ascendance and Optimization - Many suspect that the three most perfect forms of human are the embryo, the sexually mature female, and the sexually mature man. A psychologist, on the other hand, might point out that maturity may occur at different points, and it depends on what counts as 'optimization' for the individual. The most mature organisms, in this view, are the organisms that are most 'optimized', instead of the ones that fit certain criteria of physical growth and development. For example, if someone is ill throughout his or her life, optimization is

48

the time during which he or she has self-expression with medical staff, or perhaps even a time during which dreaming or some other sort of self-awareness occurred. Apparently, only athletes can be judged in terms of physical development, unless the criteria are purely superficial. Thus, it is hard to realize what is meant by an organism's 'ascendance' beyond prior conditions of development. Ascendance brings images of cyborgs, Buddhas, and aternate realities which may be as much subject to disillusionment or hallucinations as to actual ascendance. Therefore, what is meant by ascendance amounts to several things:

1. Physical / technical prowess.
2. Mental / psychic development.
3. Higher-dimensional existence.

As limits are found for physical and psychic development, dimensionalism becomes an increasingly likely candidate for genuine ascendance. It may take place through some sort of individualized process similar to individuation (Jung's term), in which the individual increasingly disassociates from prior contexts of reality, and allies his or her mind with a new context that is partly mentally constructed. Alternately, there may also be a physical model of ascendance, similar to the eat-fish-and-do-yoga tradition. The fundamental problem with ascendance appears to be what is called the 'Euclidean limit', which fundamentally amounts to the level of complexity (not pain or technology, but complexity), realizable. According to one theory, reality exists at every dimension conceivable, including perhaps whole and partial, irrational, or negative dimensions. In this sense, dimensions follow from simple mental and physical extension. Realities emerge by realizing available possibilities. Complexity and perfection means that some things are realized which were unexpected, and sometimes fortuitous. Another view is that reality occurs in distinct levels, which are realizations of nature's natural potential. For example, the third dimension might be the rational realization of the physically unrealizable first dimension. The twelfth dimension might be the realization of the unrealizable fourth dimension. The 8.9×10^{12} or 12^{12} dimension might be the realization of some other unrealized dimension which occurred in twelfth-dimensional reality. Whatever the case, the rule appears to be a compound of individual mental, physical, and dimensional extensions exacerbated by exceptional conditions which emerge out of contact and interaction with available influences. Biologically, dimensional or mental or physical extension may sometimes involve changing into another species of organism. Unexpected good or bad treatment may follow from a failure to realize one's own available potential, relative to one's available circumstances: 'the found law'.

Appendages for Sex - Extrapolating from known existing sex types yields at least two further types of sexual coupling for reproduction, as shown in the following diagram:

PREDICTED SEXUAL APPENDAGES

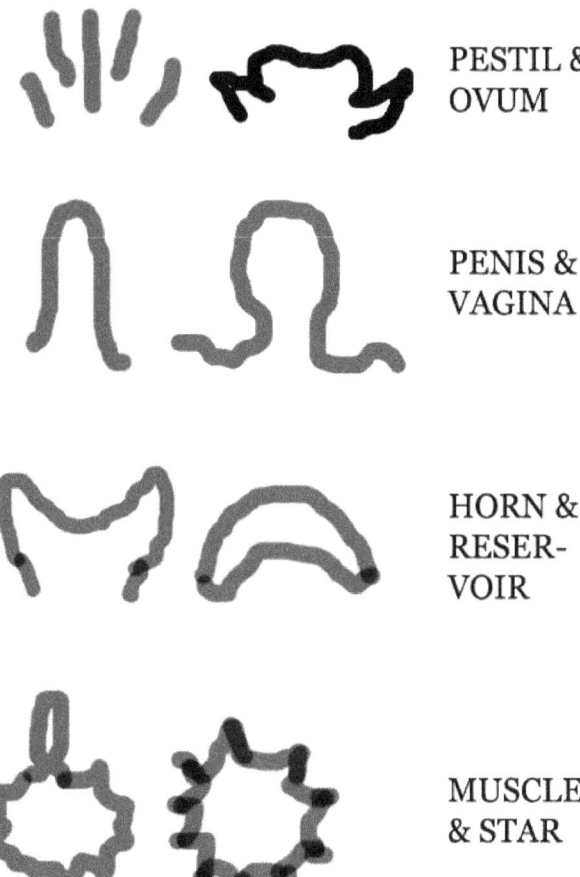

PESTIL & OVUM

PENIS & VAGINA

HORN & RESER- VOIR

MUSCLE & STAR

Assumption of the Microscope -
When the technical details of science,
alienated from a natural context, becomes
the pretext for the study of an organism or
natural event, this is called the Assumption of the Microscope. This may especially
consist of the following: (1) The assumption that a specific trait must be the only
definens, (2) The usage of 'study' as the
implied background of nature in general,
or (3) The assumption, such as due to ex-
perimentation, that an organism is the
agent of science. In extreme examples, science is mistaken for the whole of nature.
In other words, common traits of office
life, or the mind or personality of the scientist, become intermeshed with his or her
subjects; The scientist may not realize that
animals suffer, or that poets are animals.
The converse of this is not a whole recovery, and involves humanitarianism, pre-
dated by feelings of charity or humanism.
In that case, the scientist may merely feel
'educated' or 'immersed' (these become
excuses for an unconscious drive, which is
not wholly arbitrated by science, yet goes
on unchecked). The scientist does not realize that others do not have meaning for
these feelings. In the worst case, the microscope becomes a God complex
(especially via the large versus the small),
or an alienated sense in which nature is
beholden to science.

Atomic Perception - It is worth attempting to clarify by constructing dimensional biology, what is meant by the term 'dimensional', in other words, to explain how biology could be dimensional in the first place. It may be noted that phenomenologically, the dimensional aspect may depend on perception---for how else is an organism to have dimensions? Surely an organism is not merely separate faculties such as eating, excreting, or, as in some psychological theories, exclusively reproductive. Indeed, my concept of dimensionism requires a unifier such as perception as a groundwork for multiple categories. With that stated, there is a problem of reductivism, as one philosopher of biology (Hull) has stated. Simply put, organisms are not just themselves, nor are they just atoms, yet it is uncomfortable simply to approach them as intermediate processes. The reliance on any one schema: topographic, microscopic, or intermediate, would result in an incomplete concept of the organism. It might even be subject to the kind of criticism which has been leveraged against mathematics (vis. incompleteness). I have found one solution to this problem which is, in a simple set of data, to assume that atoms play a large part in the structural function of organs and organisms. After all, if they did not, organs and organisms could not have profound functions. And after all, in the context of atoms, they certainly do. I found this argument undeniable. So the individual atoms have profound functions, or else such atoms may be termed insignificant. And an organ is not made of insignificant atoms. That is the principle. It has also been observed frequently that human gametes take shapes which to each gender represent their pre-disposed dispositions for sex, such as sexual fetishes or stimulants for orgasm. To the male for example, it frequently resembles the female breasts. In the female, the gamete often represents inner chemical processes, such as progesterone and estrogen. The insertion of the penis comes to represent the repetitive process of amino-acid transcription. The point being, that these are relatively small atomic processes which have great importance for biology. The inflated importance of individual atoms goes a long way towards explaining the dimensional aspect of biology, namely the role of small processes in explaining the very large. Organisms clearly are evolved to respond to specific stimulations, which reflect their own optimal capacities in regards to nature or civilization. These aspects are the dimensional aspects of the living organism. And a key aspect of this, whether or not it is fully conscious, is atomic perception, as reflected in the sexualization of gametes, and the necessary importance of *some* atoms for biological function.

Atunement - I am not speaking of spiritual atunement but rather, something more mundane, and yet extraordinary. It may be that animals with brains engage in a special tuning process when encountering objects or situations that interest them. The atunement process allows the individual of the species to tap into a wider fund of knowledge, whether it is by memory, honed instinct, or genetic disposition. Although this is a kind of higher-functional skill, which varies in its sophistication, many species have it, and it often proves useful. Species tap into it to defend against enemies or to have a big idea. Indeed, it seems to be one of the most universal skills of nature, and goes far in explaining how even animals with small brains are well-adapted. Just as it varies in its complexity, however, it is also much more available to the developed brain, and consequently the brain may be more apt to rely on it. This effectually explains how human populations display great gifts, yet fail in the majority. It explains this by demonstrating that there is a hidden simplicity. Great accomplishments thus reflect the novelty of the entire underlying structure, much as individuals fail to be laws of nature. The greatest accomplishment of biological humanity continues to be the expression of the entire society by the individual, as expressed by atunement.

Authentic Contiguity - is an underestimated discipline in biology. Namely, the
accurate correspondence between one
characteristic phenotype and the next.
Such a characteristic need not work directly through genetics---although clearly
it is codified that way, and in a complex
view it is open to that interpretation. Instead, there may be one-to-one correspondences between characteristics such as
blood and hair, or between frizzy hair and
complexion. These need not be automatic
correlations if there is a type of modifier
involved. Instead, what may be implicated
is a kind of 'proneness' (notably also a
form of simplification), or something more
complicated like a causal routine which
adopts certain properties in accordance
with the exact biological configuration. Authentic contiguity may be abbreviated to
the classification {a, m, c}, where 'a' is an
attribute, 'm' is a modifier, and 'c' is a
causal reaction. The total can be represented as a complex number in which the
variables are considered independent.
Thus, unless there is overlap between the
types of calculations, the classification is
unique for every *comparative* relation between one attribute and another (because
they have different responses unless they
are grouped together). The values however, are not unique for organisms where
attributes are not compared. Similar numbers emerge for example, when 'm' or 'c'
are removed. This assumes that each variable has a quasi-exponential formula:

Axotic Development - *Typological Method.*
Axotic: By / relating to axes.

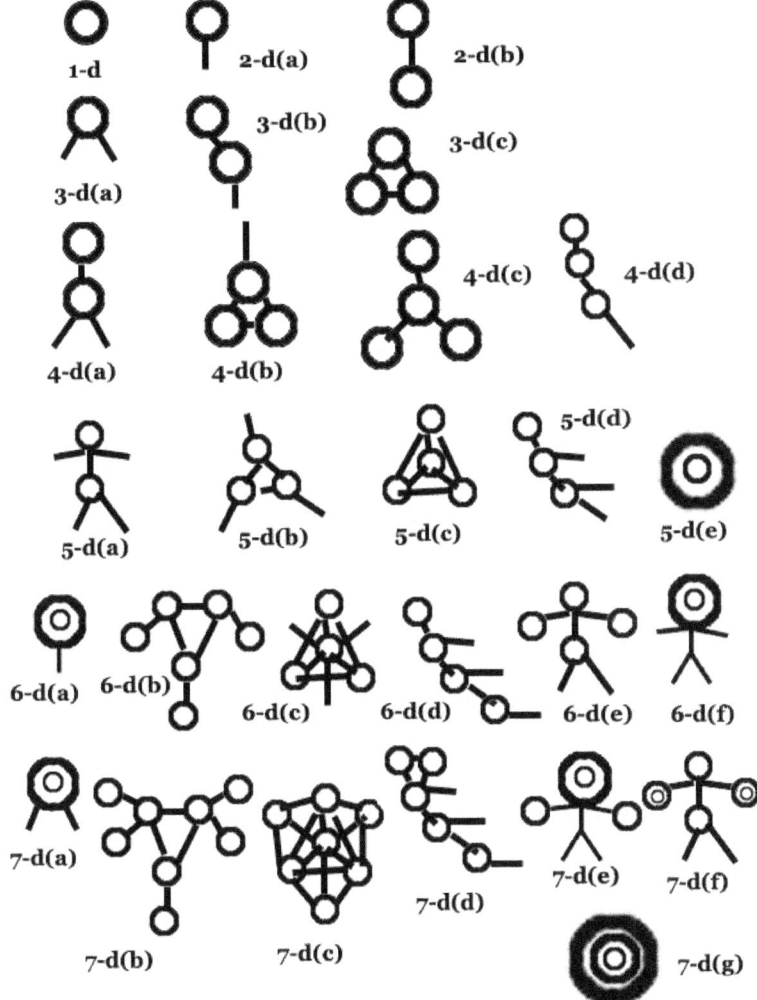

Here can be seen a variety of developments, such as the early incidence of genitals, the early incidence of madness, and the variety of bizarre structures which compete with or complement human beings. The spheres are accesses to new dimensions, such as higher consciousness, or use of tools. Independent of other hierarchies.

============ **[B]** ============

Barr Body - (female cell anatomy), One of the many intelligent organs of a woman's body. The Barr Body stores de-activated X-chromosomes in condensed form throughout the female body. It is even possible that there is connectivity which predates sexual intercourse, implicating a 'vertical' functionality of the female genome in tandem with X-chromosome synthesization. The purpose of this structure is essentially genetic in origin, a chemical rather than survival adaptation. It may assist with processing functions such as social judgment which males do not have, and does so in explicit relation to sexual or emotional arousal. This is one of the major genetic dispositions of women to express the phrase 'thinking with the body'. Such an organ could certainly be compared with the brain, and in this role the function may be associative or synergistic. It is only for science to determine the exact nature of these mysteries. The secondary role of the Barr Body probably relates to early fetal development, and its genetic character may be determinate of features such as hetero versus homo sexuality, nymphomania, and lying.

Binomial Nomeclature - The formula, such as *Tabanus opacus** or *Homo sapiens*, that defines plants or animals or other possible types (e.g. tropophiles, heliocytes), into genus and species levels, indicated by a first-capitalized and first-lowercase word, respectively. Italics are granted exclusively to official, or in the case of philosophy, demonstrative, cases. Trinomial nomenclature, using three words, is commonly used to indicate the sub-species of a species, for example, *Homo sapiens* sapiens ("thinking man thinking").

*Tabanus opacus is the horsefly.

Bio-Hazards and Other Bromides - Although I would like to take a psychological approach to these exceptional influences, some factors stand in the way: (1) When they may cause death, (2) When these things may cause genetic mutations or other species-level changes, (3) When these things may cause other non-genetic changes, such as perceptual shifts and non-adapted changes, and (4) When these things are of irrational origin. Although I think irrational causes can be overcome by reason or genius or alternately a Utilitarian moral, and non-adapted changes are either non-serious or may later incur adaptations, and genetic changes can be interpreted as responses to optimization, the justification of change by death is only justifiable under a psychological theory, or perhaps an economic one. One wants to believe that death is contextualized by thoughts and actions, or else it becomes irrational, or under an economic theory, life may be fractionally irrational. So far as psychology goes, survival is a kind of affectation. Other major effects on the individual---sexual attraction, aesthetic repulsion, spiritual abstention---are consequences of exterior affects upon the person. Thus, following a Darwinistic species-influence model, the role of psychology in biology is primarily *affective*, and may relate to causes or networks. It may be seen that this is the ultimate purport of bio-hazards and other bromides, since from a survival point-of-view, such things are genetic synthesizations of physical experiences which are themselves synthesizations of the sensitive (qua psychological) outer-core of experience. Thus, the primary genetic response to bio-hazards occurs first through psychology, and only secondly through direct physical interaction, although *sub specie parens*, the psychological is also a direct product of experiences.

Biological Coincidence [an attempt to re-habilitate an abused concept] - Where psychologically a coincidence is potentially meaningful, such as unresolved emotions which do not feel judged, or an unintended play on words, in biology coincidence, by any adequate standard, has definite meaning, because it is always an association between two very real conditions of nature. (A true biological coincidence, in the traditional sense, is something being ignored). Consider, for example, that the length of a word someone is comfortable reading may be similar to the distance between the reader's eyes. India, known for its long names, is also known for a large space between the eyes. This quality has even been seen as seductive. The distance an ant must crawl to reach the end of a pencil is also a biological coincidence. So too is the co-evolution of different species in response to similar terrain, and the way plants do not tend to grow in outer space. But if some plants did grow there, their survival would also constitute a form of biological coincidence: the availability of adequate circumstances for survival. In every case, there is a relativity only by the relation of a population to an application, and there is also simultaneously a referrant to one definite fact: the outer world. If there is a modification of the definition of biological coincidence, we are likely to accept it in these terms.

Biological Conflict - As we enter the machine age, which I associate with the cusp of the fourth dimension, it becomes important to distinguish conflict from more egalitarian concepts such as the food pyramid. Food is no longer the only source of conflict. Many forms of combat occur because of psychological anger, or, even in the case of simpler animals than humans, in the form of callous aggressions which do not serve any practical purpose. This is illustrated for example, in the case of a cat playing with a dead vole which it offers to its master instead of eating itself, or in the case of lions who will not eat their dead opponent, presumably out of disdain for that animal. As conflict has progressed with the success of the human species, the major factor to notice has been the success of intelligent armies. This has carried over in other species as well, such as ants, who have an entire soldier class, and birds, which in the case of weaker birds, will often fly in groups in order to combat larger birds. The paradigm of area-of-effect has made birds less-than-invulnerable to fights with humans, instead being made vulnerable only when humans become angry, or as a result of secondary effects like lack of food, or forms of environmental pollution. Environmental pollution has served to stave off the human paranoia that some other species will begin to dominate, raising visions of 'mutant' species who somehow overcome genetic damage in order to become dangerous contenders. Another factor has been the objectification of conflict, through the creation of weaponry. Earlier, weapons were considered necessary an extension of psychology, such as a bird's claws, or a tiger's teeth. With the advent of impersonalized weapons (I will say a gun is at least slightly impersonalized, like a mask), it became possible to see warfare as something which occurred independent of instincts, and independent of speciation. A vision was aroused of a world full of traps and dangers, a place where a robot could step in and make a dangerous arbitrary decision. As conflict reaches the fourth dimension, however, a new factor emerges: armies must be *interesting*, or else the intelligence of the soldiers will exceed the desire to conflict. The alternative is essentially madness. This sums up the quandary of fourth-dimensional warfare as I understand it: either conflict will evolve towards pacifism (by becoming desirable, hence interesting), or soldiers will cease to evolve and thus maintain the animal patterns of the third dimension, or otherwise, conflict will begin to consist of sheer madness, and will be subject to that form of judgment. Predictably, a mixture may emerge, but as the number of dimensions increases, increasingly conflict will be a chaotic element in disparity with the level of interest evoked by the degree of complexity. In the future conflict will look something like the willingness to switch back to tube televisions from something more advanced. The need isn't there, unless there is conceptual brutality. However, as most thinkers know, the need for conceptual brutality is very small, unless the thinker is mad, or reconnecting with the primitive impulse.

Biological Deduction - Against the empirical trend in biological philosophy is the
categorical one. While the characteristics
of animals are not often termed opposite,
and therefore there may be a possible difficulty in posing exclusivity, there are many
contexts of biology which do consist of
categories. There is also the inherited categorical aspects of behavior from psychology and neuroscience. It may be noted that
oppositeness also suggests opposite loca-
tions, life-conditions, and perhaps even
emotional states such as personality and
intelligence. There is thus a possibility of
modular in addition to *qua* absolute forms
of categorization. Categories include not
only formalistic *taxa* of animal and plant
species categories (when deemed exhaus-
tive), but also cycles of birth, youth, maturity, and death, and cycles of biosomal in-
terdependence, such as aspects of the carbon and nitrogen cycles, consumption and
excretion functions, etc. Continue reading
to see the logical method of deduction;
(For related logic, see Resemblancy).

CONT'D:

A METHOD OF BIOLOGICAL DEDUCTION

SET ONE

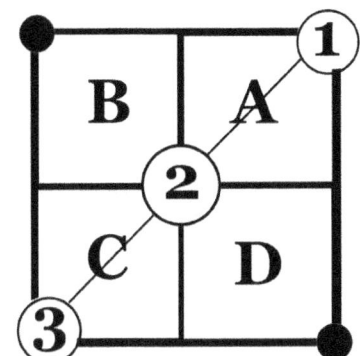

1-2-3
primary axis

ABCD
consistent
context

SET TWO

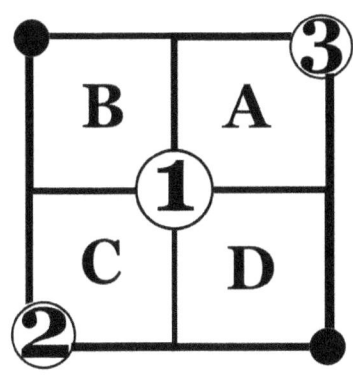

SET THREE

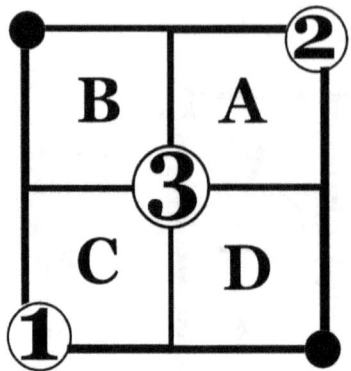

This method (Sets I-III) adopts one primary axis. A variation of this would adopt 3X additional categories to explain a context of the judgment, such as habitat, environment, or technology concerns (across B -D and also occupying the center). For most purposes this second method is more of a permutation, and thus has less efficiency in relating clear biological categories.

Here are sample sets that may be used for the primary axis:

[Interaction, Dynamic, Adaption]
[Consumption, Energy, Intelligence]
[Survival, Action, Attraction]
[Consuming, Processing, Excreting]

The sense is somewhat metaphorical when the assumed categories do not overlap, e.g. when one set refers to animals and the other to plants. There is still some validity in those cases, however, albeit more obscure. They are simply cases in which categories must be cleverly combined, and thus equated with something new, or else, something highly specific that already has examples.

Here are sample sets for the consistent context:

[Birth, Youth, Maturity, Death]
[Winter, Spring, Summer, Fall]
[Beginning, Struggle, Conflict, Resolution]
[Identity, Resource, Strategy, Immortality]
[Efficiency, Desire, Time, Discretion]
[Potential, Generation, Realization, Recovery]

Now I will generate the data using the following sets:

Primary Axis: 'Consumption, Energy, Intelligence'
Context: 'Birth, Youth, Maturity, Death'

The logic follows the following rules, generating three sets:

[*The middle position simply serves as the title of the individual section*]

[method of biological deduction:]

I. Title: 2
1A: B : 3C: D

II. Title: 1
3A : B : 2C : D

III. Title: 3
2A : B : 1C : D

[*Notes: alternatives involve altering context B-D or treating B-D as an axis by rotating the diagram clockwise; The result is logically the same, but the choice should be justified according to a linear pattern of development, or using the philosophical method of opposites*].

The results generated are as follows, according to the chosen terms:

I. Energy: Consumption of birth has youth and intelligent maturity has death;

II. Consumption: Intelligent birth has youth
and energetic maturity has death;

III. Intelligence: Energetic birth has youth and consuming maturity has death;

I find these are fairly adequate and broad-ranging, and desirable definitions, assuming the contexts are accepted to be exclusive. Other definitions could be produced in relation to other context comparisons, as well. Here is an alternative, advanced interpretation of the same method of comparison:

Nathan Coppedge

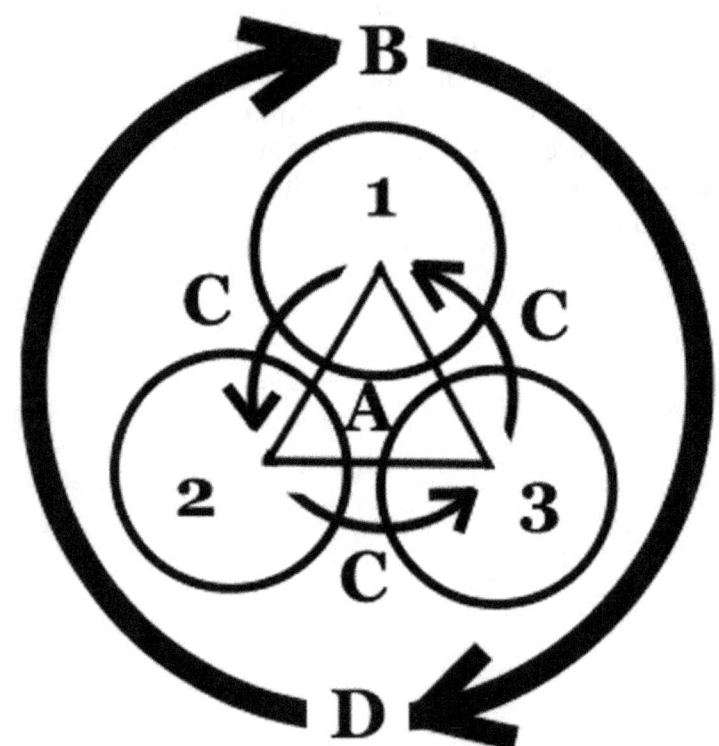

**BELOW ARE SHOWN OTHER VALID
COMBINATIONS, IN THEORY
NOT SEMANTIC: THE THEORY
IS NON-OPPOSITE WITH OPPOSITES**

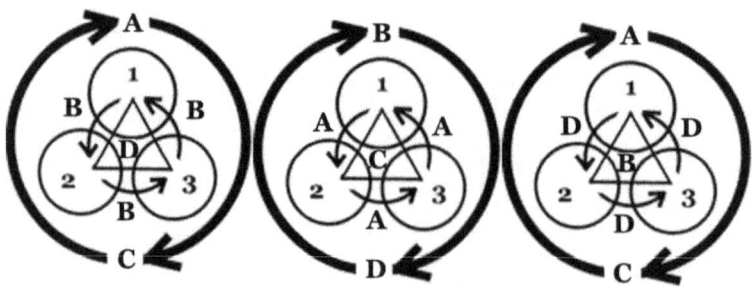

[INTERPRETATION OF ADV. BIO. DEDUCTION]

Implication on the advanced method:

These (earlier) theories concerning natural order, and corresponding theories for any remaining categories, may easily be suggested and expanded by use of this method, including the earlier format; The advanced diagram may be useful for visualizing if someone attempts to interpret the genome this way, but the deductive method remains a three-category system, since the position of inner- or outer- logics from the trinity is identical. In other words the system is logically identical, yet more complex, than the earlier method.

Biological Eschatology -

Wrinkles

Teeth Gaps

The common property of fissures or gaps in the body appears to be the loss of some element, whether it is a tooth, or time, or bodily waste. Even the hands' fingers and the feet's toes seem to represent a potential loss: the loss of held objects or modes of relationship such as family ("tied hands"), or location ("rooted feet"). Perhaps this paradigm is as basic as the need to eat plants.

Biological Imperatives -

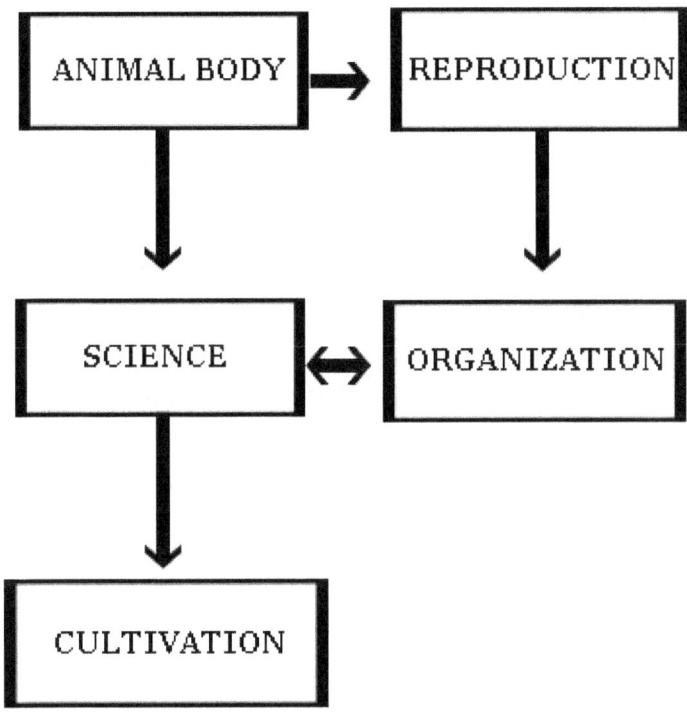

Exception to the above are group regression and ignominy.

The above gives some flexibility, thus 'protecting the young' could be seen as a function of the body, (social) organization, or even science or cultivation. Various degrees may exist, which may or may not possess a functional relationship.

Biological Lens -

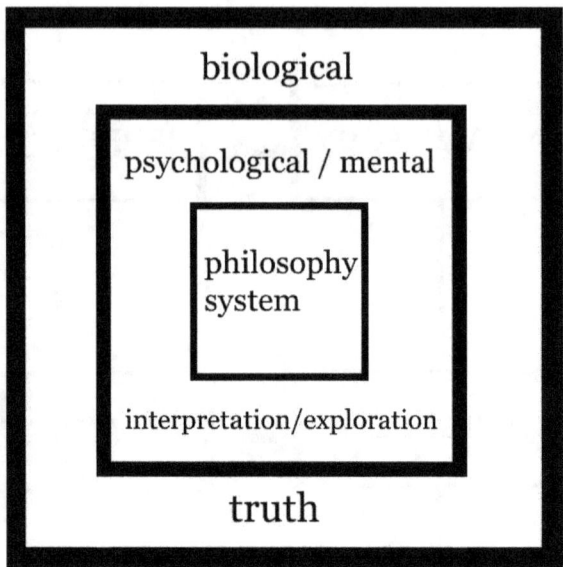

::

Biological Substrates - Substrates ---
layers of substances, such as sediments,
meshes, moss patches, and levels of
growth --- have a multi-partite relationship to concepts of organics in matter. In
the case of a moss patch, the substrate
layer is little more than a quantifier, and
also serves as the organism's datum of existence. It is easy to see how such a layer
could simplify the dimensions necessary
for survival --- e.g. by reacting to gravity
against a wall, reproduction is restricted to
horizontal off-shoots, and health occupies

a short vertical. Thus, where organisms have senses, such a format allows the organism to perform some basic comparative functions against the status quo --- a form of self-selection. In more artificial forms, such as meshes, the location in the 'grid' may determine a plant's ability to reproduce or be watered. Major events occur between immediately neighboring plants, which must either be symbiotic or compete. As gardeners know, the grouping of multiple plants of one type in one area can increase the chance of the plants' being noticed and hence getting watered, or being adjusted to the sunlight. In the case of sediments, altitude affects the availability of light and other resources (such as dampness, or mineralization), however, organisms may often have the ability to grow at multiple levels of the sediment, as we see with root systems and human dwellings. It is that level of variation which provides a categorical platform for the properties of speciezation, leading to unique, often overlapping environmental niches. In essence, organisms learn to thrive that can cover space, and thus in optimal cases inhabit a 'dimension' of survival (the vertical).

Biology and Time Travel— The phases of the seasons and changes of termperature offer one prospect for recording travel that occurs outside of standard time. Other prospects are measures of bizarity such as extreme beauty or extreme longevity: perhaps these types of persons have tasted something of the immortal, through a process called 'conformation to the archetype'. The primary theory must be that immortals and time travelers are a different type of animal, perhaps given to certain types of creativity, or certain types of liberated genius. Account must also be made of how the time travel event could play a role in sexual and social chemistry. The traveler may be a pariah who has a kind of anachronistic syndrome, a sense of not belonging in his own time. In despearation the time traveler may feel he belongs in the future time as much as the past, creating a break with reality and an 'inconsequation of the present'.

Biology in the Humanities - An approach to biology in the humanities.

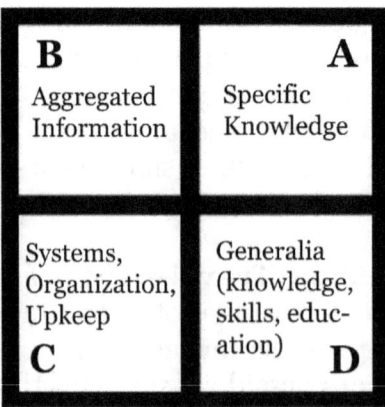

B Aggregated Information	A Specific Knowledge
Systems, Organization, Upkeep C	Generalia (knowledge, skills, education) D

I have been trying to ferret out exactly

what matters in biology, and the result inevitably favors specifics while catering to generalities. There is room for systems biology, just as there is room for aggregated information. The four categories favor knowledge and information which may be environmentally practical, have scientific applications, or serve to stimulate public interest. Aggregate information in particular is endlessly applicable to the academic biologist.

Biotechnology -

Basic Biotechnology

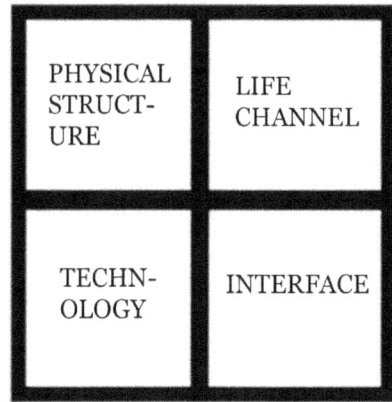

1. When the technology is the physical structure, the life-channel is interface. 2. When technology is the interface, the life-channel is the physical structure. Essentially, the choice is life or technology, or the only choice is a combination. If the

question is invasiveness, technological interface may be differentiated from technological structure, and living within interface provides an alternative to living channels. Life becomes the vector for technology that serves technology.

Bodily Nightmares - occur in the natural body, especially in response to poisons and unusual odors, hallucinations, or other strange perceptions. Essentially this goes to show that the brain is united with the real body, and that aspects of bodily chemistry reflect mental chemistry as well. Bodily nightmares may be studied as concerning or almost occurring inside, the neck, heart, feet, or even genitals. One view is that these are merely mental associations for these areas, but the nightmares occur in real time, so there is no denying that the localization is naturalized. Another view is that these sensations are associated with the primitive brain, however, many researchers have in turn associated the primitive brain with the most vital and complex or unconscious urges of human development. Whether or not a physical location is found for these sensations outside of the brain, it is undeniable that the locatedness of these phenomena, whether they are associations or real physico-mental apparatuses, provides a useful extension of mental cognition that might otherwise be assumed redundant or irrelevant.

Body - Categories of:

1. Animals that survive in a suspension, such as microorganisms.

2. Animal that survives by filling an early niche. Many of these are very small, for example, aphids. Some of these share some properties with plants such as plankton.
3. Small, subsistence animal, such as an insect or small fish.

4. Large predator (ocean or land). Predators often share similar properties, such as a large jaw and body maximized for the amount of available prey.

5. Large, vegetative body. Cows, horses, and whales, for instance.

6. Small body designed for dexterity. Mice, chipmunks, gerbils, etc.

7. Domesticated animal (larger brain). Cats and dogs.

8. Hominid designed for brute survival (up to early humans).

9. Hominids designed for indoor living and intelligence. Humans and similar species. Alternately, a hominid designed for similarly advanced living of a different type (say, on a fire-y planet).

10. Small, specialized body (in this case more advanced than a human in some respect, such as having 'arcane intelligence,' greater technical motor skills, greater survival rate, etc. Different evolutionary origin, but may develop in tandem with the human).

11. Any organism adapted to supplement the intelligence of a super-computer. This type includes many varieties, from engineers to psychics, to entertainers. This species is marked off from the next by its lack of immortality.

12. Beings that are immortal with the exception of bizarre accidents. This requires some degree of profound intelligence above the other types, but also depends on circumstance. Amongst this type, it may be said that only the best are fully characteristic of the species. However, that is not to say that the characteristics of the species are not advantageous to earlier types. The peculiar characteristic of this type is to make immortality a way of life.

13. Further types may be had by adapting to the specific environmental character of immortality, resulting in a 'new animal'. Perhaps this type could 'survive death,' teleport with its body, turn invisible, etc. Aspects of this type border on the superhuman, beyond mere mental powers. Various types can be discerned, depending on its mode of living. Some forms may be highly

dexterous, whereas others may be seden-
tary.

Brain Centers -

Distributed
Centers must
depend on some
falling substance,
like rain or heat.

Heads seem to
perfect the idea
of an efficient,
Mobile Center.

But, what if the
upper body had
multiple brains?
Each would need
to be distributed
above points of
balance...

Finally, the
brain is extend-
ed into nerve
endings in other
appendages.

See also under Distributed Brains.

Brain Concept Paradox - few biologists
doubt that the human brain is a physical
construct and that it is meant to function;
But what if some or all brains are dysfunctional? This would open the door to problems more conventional to philosophy, e.g.

is the brain the only form of function, and is the only absolute function a simulation? The problem alienates some behavioral functions in biology by asking the question 'what if, in spite of all appearances, the brain is dysfunctional?' Perhaps then the brain is the only context for functionalism. This raises the further question that function may be purely individualistic. The problem is similar to the philosophical problem of the brain-in-the-vat. How could we know the difference? What is the functional difference without knowledge? Should we respond to or study unknowns in biology? One solution is that there is a similarity between cognitive and behavioral systems. But doesn't this de-materialize the biological idea? A starting point is to realize that thought dynamically effects the behavioral role of materials. (Biological function may have a different significance in the context of cognitive function). In some manner or other, thought may be more concrete, more conclusive, than some properties of matter. In this way, behavior is a function of the mind, not the reverse. If there is middle ground, perhaps it is not a minimal organization, but instead a function of mental and environmental contingencies. Perhaps, for example, individuals are more prone to respond to the unknown than to something they understand, much in the tradition of a gossiper. Theories such as mind-brain identity theory depend on an understanding of to what degree the brain depends on the environment, and vice versa. The problem grows more complex when there are varying types of organisms, and varying degrees of mental development. The simplest answer perhaps, is the answer I have already hinted at, which is that 'definitive events' such as thought and nuclear fusion serve an authoritative role for matter, in defining the character of the interaction. Thus, thought has a one-up advantage, in being one-part experience, and one-part definition. Of course, it is easy to reject this theory, but the result is to favor a form of concretism, which sets a limit upon the complexity of life. One can imagine four categories produced from the two scenarios: [1] A stupid human (concretism), [2] Flowering tree (thought-definitism), [3] Intelligent human (thought -definitism), [4] Living flames (concretism). It is easy to see how concreteness produces desirable categories, but fails to bridge the somatic gap, the solution to which promises to explain the mind-brain identity theory.

Brain Function -

The human brain has two hemispheres (right and left), plus a few coordination organs, which span in-between. The right hemisphere is the associative and coordinating brain, and is also related to artistic and process-related functions. Strong emotions may be located here. The left hemisphere is the deliberate, task-oriented brain, and has been associated with mathematics, music, and metaphor. Returning to coordinating functions, the Hippocampus is the largest of these, is sometimes significantly large, and is located near the back and base of the brain, just in front of the medula. It is responsible for making sure that the brain is 'doing something,' and is related to engagement and emotional sensibility. The Thalamus is the largest nearby coordinating organ, and essentially integrates between the brain and the rest of the body. Also related to these is the Hypothalamus. The Hypothalamus is important because it coordinates the brain's feelings of normalcy. It is therefore integral to the brain's functioning, by supporting the functions performed elsewhere in the brain. It is located in front of the Hippocampus, approximately in the center of the brain. An impaired Hypothalamus is sometimes associated with damage to the temporal lobe.

Next, the temporal lobe, which is part of the cerebrum (and which we will say is not usually damaged, and is often larger than the thalamus), is related to time-based functions, manual dexterity, and general body-consciousness. For most people, the temporal lobe performs some important functions. The medula or visual cortex, located in the back and bottom of the brain, is related to the intensity of mental imagery, and the ability to describe things in detail. Many artists have a large or complex visual cortex. Next, the cerebellum is in the back upper part of the brain, and is associated with bodily grace, ambidexterity, technical crafts, and other advanced motor functions. It is also related to technical aspects of memory, such as recognizing an image. Lastly, and most importantly, the Cerebrum is the largest part of the brain and is most directly related to intellectual and language functions. Having a large and / or complex cerebrum virtual guarantees an interest in intellectual activity. The frontal lobe, which is located at the very front of the Cerebrum, is related to intellectual balance and ideology. It has sometimes been associated with memory functions, but it would be more accurate to say that it controls 'higher processing'. With support from the medula, which is another name for the visual cortex, and the cerebellum, the frontal lobe provides subjects for the temporal lobe, which are then translated via the thalamus and hypothalamus (including sensory input), into a cogent and coherent concept of thought. Sometimes the idea of thinking is unconscious, and merely serves to accentuate outer experiences. However, as intellectual ability improves and develops, self-concept depends on some conscious knowledge of the brain to make forwards leaps. Some of this knowledge can come directly through experience, whereas other parts of it are more likely to be acquired through research and study. In past history, these leaps were ersatz and over-wrought as they were rare, but in more modern history, scientific knowledge has allowed for instantaneous acquisition of real bodies of knowledge abbreviated by the human or post-human brain structure. See Also: Metaphysics of the Brain.

Butterfly Effect - The effect popularized by numerous popular books, including Andy Andrews' *The Butterfly Effect*, and works in physics and economics such as James Gleick's *Chaos: Making of a New Science* and Nicholas Taleb's *The Black Swan*. Essentially, there may be an informational and exponential relationship between one condition or set of conditions and the next, creating complex strings of reactions in which complex predications 'prey' on the unknown, resulting in events which may be as global as they are bizarre.

A related theory is that of *Amoebic Assimilation*, in which it can be anticipated that if there were once, or were ever, a universal life form (such as say, an amoeba) then there is a strong likelihood that, whether we know it or not, all other forms of life are derivatives of basic or complex properties or conditions of that organism. Not just in a speciezation sense, but in a physical, environmental, and metabolic sense. However, that may depend on some predicate of a hidden reality in which impressions of life forms persist intensively. If all species are merely impressions of some separate hidden logic which cannot be called an organism, then amoebic assimilation disappears. So not only may it only exist in the case of a single unified species, but it may also only exist in the case of conceiving that all matter is biological. This opens the thesis to an intriguing sense of virtualism. However, the properties of reality defining the case divide it into two schools---'fossil realists' and 'fossil formalists.' Formalists believe that life is like a syntax which prefigures all forms of matter. Realists believe that all organisms are parasitically linked, and thus the relation to non-biological matter is more desperate and antedevulian::

===========**[C]**===========

Categorical Adaptation - Is often seen
in absurd terms in the human era. A child
highly familiar with visiting hotels may
feel that they are his or her only natural
environment. When asked what should be
improved about the world, he or she may
reply that 'everything should be made like
hotels'. Yet the advantage may be arbitrary. It may not pose an inherent survival
benefit. Indeed, the hotel may even alien-
ate the individual from future mating
situations and education, or reduce his or
her motivation to compete for resources.
The case is much like a swallow who develops larger wings, or a turtle who grows a
harder shell. The advantage served is ambiguous outside of a specific context, but
creates an attachment by the individual.
Resources may have been spent, risks may
have been taken, and the organism needs
the advantage to pay off. Accepting the
fullest advantageousness of these claims to
survival is almost like believing in God.
One may ask, what provides for the condition that an advantage concerns only a single category? Is such a condition a function
of the organism, or instead the environment? What if the advantage is more intelligent than it appears? Certainly it is possible that the new factor (in the case of the
swallow, the turtle, or the human) is the

last hurdle in a gradual improvement that has finally reached its apogee of perfection. But humans look on other animals with rightful criticism. And so, people should also be critical of themselves. Then, it can be said that the development amounts to a kind of thesis of adaptation:

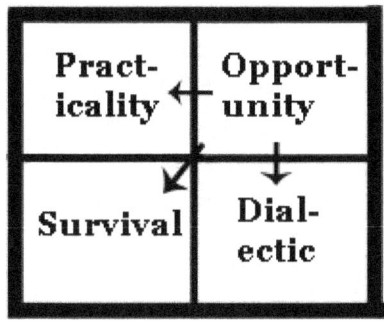

These categories serve a wide variety of functions (reading in standard categorical fashion from upper right counterclockwise):

A. Motion, lifespan, features.
B. Foraging, behavior, use of objects.
C. Appeal, dominance, attraction.
D. Symbiosis, intelligence, conformity.

The overall effect is to describe the migration of specific 'measured' opportunities such as lifespan and motion ability, into general survival, such as foraging, attraction to mates, and display of intelligence.

Nathan Coppedge

Categorical Dilemma in Biology - It does not seem fair to divide the globe into computational categories, still less to axiomatize that stars cohere to some plan of opposite extension. As appealing as these theories were at the time of Paracelsus, they do not hold the same weight today in relation to what may be called the empirical realm. The lemma of this dilemma is that some method of categorization may still be possible. Indeed, the lack of obvious categorization may be a product of latent complexity. Nor does the absence of organization concepts in the world suggest that animals or plants could not be conceptualized, only that some of the concepts are outwardly disordered in appearance. This is the kind of problem that has been common in economics, that while organization is not obvious, it is not that it is entirely undesirable. Consider that biology does not entirely eliminate opposites, which I have found to be the most convenient form of deductive measurement. And while many animals are not themselves 'opposite' to any species, it may often be convenient to assume that they are. Indeed, individual characteristics can be opposite even when the overall organism is not. The question becomes, what is the difference between a biological and a philosophical or psychological method. I think the answer is that in biology, there is synergy, whereas in philosophy there was a unity, and in psychology there was a differentiation. The synergy provides an opportunity to apply organization to biology.

Organization decides, in a variabilistic-*qualia* manner, that every morphization of the individual results in a standard type, which further defines standard correspondences, creating standard phyla for comparison. Thus, organization becomes a key to morphization, the synergy into ever more specialized or ever more perfect formulations. At this point, the philosopher would speak up and say, even generality is a form of specialization. And, indeed that is the case. But instead of the coherent mentality of philosophy, biology proposes animals---entire organisms---that oppose other organisms by single lemmas, like base consumption, holistic competitiveness, or special advantages. Organization proposes that the system of correspondences is as standard, and as evolving, as the animals themselves. The lemma at this point is that the very nature of opposition is continually prone to modification, and thus, the functionality of individuals is radically dependent on the functional nature of society as a whole, in such a manner that entire social ideas can be amalgamated as small parts of a very functional organism. In other words, biology is more socially than individually exponential. By contrast, psychology laid the claim that all things are psychological, and philosophy laid the claim that philosophy is the rudimentary system. In biology the 'rudimentary' is hacked, and psychology sometimes loses its definition. Indeed, consequently, it is as though biology fractures philosophy, and quantizes psychology. As such, philosophy becomes the object discovered at random, whereas psychology is its exponent. The ultimate theories ultimately twist and turn, until the object discovered at random is, by fiat, the most significant thing, whereas psychology, which is at the root of everything, disappears on the ultimate level. In this way, the roles ultimately reverse::

Categorical Genotyping [Genii] -

It is important to realize the efficacy of this system. The method described in the following diagram is made to integrate with Clausal Development methods. Ostensibly, this would allow for a balanced approach to artificial, intermediate, and primitive forms of life, not by type-exclusion or observable organization, but instead by classifications of naturally realizable functions and meta-functions. Here is the diagram, in which it can be seen that the four categories are functions of one another:

Mode System	Genotype System
Function System	Adaptive System

Genotype is a predominately indexing function, allowing insight into sub-causal relationships. Function, in reverse of genotype, is a predominately associative type that spans biological boundaries. Mode is the entity unifier of associative functions. Adaptation allows some miscellaneous typing, across associative and entity boundaries. Function allows a strong approach to genus-typing, while adaptation becomes an extreme answer to modality approaches.

Chemical Cycle in Biology -

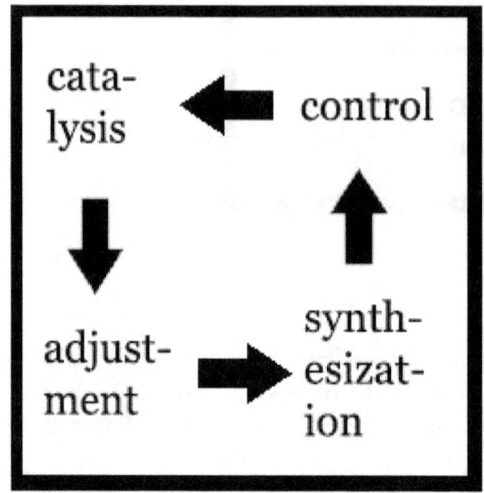

Stage I (Control): During this stage the plant or animal develops cells, organs, etc. to control substances.

Stage II (Catalysis): During this stage the organism may develop unique substances to aid chemical processes. These may include some things as advanced as genetics, or as simple as imbibed substances.

Stage III (Adjustment): During this stage the organism evolves to better adapt to its own chemistry.

Stage IV (Synthesization): During this stage the organism is sufficiently advanced to create unique compounds specifically for its own individual purposes. This may be one or many substances, with many functions.

Clausal Development [philosophy of genetics]-

Abverse clausal development: This is also called the causal-clausal view, under which (1) Categories develop independently, (2) Each category develops from a prior category or combination of categories, (3) Developments occur by material implication, and (4) Complex theories of development imply explanations of material implication or entailment. A predominately material and un-complex view, it notably prefers correspondence over coherence as the basis for the theory, creating some conflict with metaphysical viewpoints. Theories of robotics and nanotechnology seem to thrive under this theory, because theories become a function of development and developmental choices. Choices then become a sophisticated plug-in theory for the function of these entities (*a posteriori*). It does not become necessary to explain nature as a whole, unless there is some conflict between not only individual entities, but between theories of these entities. And theories are more likely to function on the basis of common ground. An alternate view, called *nominal clausal development*, is the ontological explanation of development, which occurs by fulfilling categories. This theory has often been ignored because of the apparently teleological function of nature. However, in the information age, object-oriented ontology has brought new interest to bear on category-fulfillment theories.

Coherent Theory of Clausality

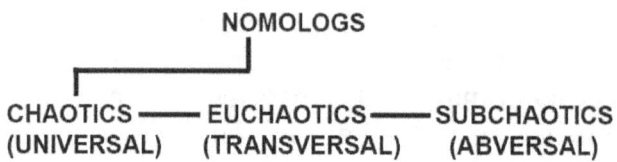

NOMOLOGS

CHAOTICS —— EUCHAOTICS —— SUBCHAOTICS
(UNIVERSAL)   (TRANSVERSAL)   (ABVERSAL)

The four types divide into four types of clausal development, which reflect different levels of assumptions about biological reality. The four types are Nominal Clausal Development, Universal, Transversal, and Abversal. The chaotic type can also be called *Prolifae metaphysica*; Euchaotics (distinguished in this case from the system which uses the much more conventional type called the Eukaryotes) are all presently known biological types, presenting some differences between plants and animals. Subchaotic is the emerging robotic form of life, and any form of life analogous to that, for example, heliophiles that develop to feed specifically on spaceship exhaust. The need for a great deal of variety within a given tier of the system is emphasized by the emerging age of *schizoriginal nomenclature.*

Concealed Functions - Not all concepts of concealed functions hold that functions are actually concealed. Thus I introduce two unconcealed variations, followed by the primary example of concealed function, called concealed information.

The first approach is that these secondary functions are merely a function of the desire of the observer (interpretively or functionally). The result is a purely psychological approach, a regression in my system. This might treat real functions as relative or correspondent functions, and treat interpretive functions as functions of applications. When applications can be treated as tools that are a function of biology holistically, then interpretation becomes both psychological and biological: it has justice, but also refers to something.

A second approach is chemical and cellular, but may not apply to all types of organisms. In this approach, concealed functions are always a function of the overall function or dysfunction of an organism. For example, chemical pathways, if they are a concealed function, are likely to be a result of a centralized brain, if the brain is centralized, and a de-centralized nervous system, if the nervous system is de-centralized (hypothetically).

A third approach is the concealed information approach, in which processes may be invisible for on reason or another. This flies in the face of simple assumptions in biology. In this case, however, concealed functions can be treated as a function of concealment, creating a firm concept of the inner and outer organism, with implications for all inner and outer concepts of biology, including behavior and intelligence::

Conceptual Advances in Biology - The most commonly cited advance in biology since chimpanzees is frequently cited to be technology in general, including both weapons, manufacturing implements, computers, and other objects, including their miniaturizations. The main factors of biological advancement continue to be sophistication, integration, and globalization. In recent history, it has been realized that sometimes the concept may even involve a process that affects an entire planetoid, planet, or star, such as the Dyson Sphere theorized and designed to extract energy from an entire star. Another such concept is the idea of an entire planet that has been turned into a living brain. Most advances, however, are projected to relate to a combination of integration and globalization, specifically the economic popularity of products designed to enhance existence through the use of technology. Increasingly these products will relate with information and also with the specific experiences and advantages that benefit the people that use them.

Continual Evolution - There is no better example of continued evolution than that of the Wels catfish, a fish that has adapted to snatch pigeons that land near the lip of the water. Examples may be rare, but occasionally --- perhaps in the most notable species --- adaptations are made in response to extreme environmental modifications. The incidence of quasi-domestic animals such as the pigeon---or perhaps originally dragon-flies, is perhaps what led the Wels catfish to begin its gravitation towards land. In other cases it can be observed that ants prefer to build their mounds underneath frequented picnic tables, an extension of the concept of building mounds near their food sources. It is even possible that Ivy-league students have developed a high incidence of mental illness to avoid the military draft --- a case that occurs less and less the more political influence the students possess. It is interesting to see that in countries that are less capitalistic, the elite status still corresponds with political status but not always money. Where money appears, money begins to be the factor which controls influence. But not always. Similarly, military families are known to be more likely to suffer from wife- or husband- abuse, and child abuse, and are more accepting of this phenomenon, which may be an adaptation designed to curb fornication, and thus to prevent venereal diseases under the view that these groups grow up to be sexually active. Not all attempts at evolution appear

successful. Indeed, if an animal has a perspective on his or her adaptations, it is not
likely to be 100% positive. Most major adaptations occur in response to stress, and
still involve handling much of the stress
that was originally present, sometimes by
drastic pro-active measures, or re-
allocation of mental resources. In the case
of the Lionsfish, which is growing more
numerous in the world's oceans, the response comes as often happens, out of climate change, and the hard-as-would-have-
it ability to produce venomous spines
which render the fish invulnerable to
nearly all predators. It is possible to see
the Lionfish adapting the crawl on land as
its supremacy becomes even more unques-
tioned, although humans already have the
ability to destroy them (with some difficulty). It is the delicate balance in the first
venturing onto land which captivates human audiences about the Wels catfish. Perhaps the novelty of welcoming another
creature onto land will allow this catfish to
thrive and grow, ultimately beside the human populations. In the long scheme of
geographic history, no theory is completely
absurd. There is not even a requirement to
always be practical. We have to remember
that the pressures forced upon us by Dar-
winian survival are largely artificial: in
terms of separating categories, they depend on networks of interactions which
are not originally competitive. When looking for new forms of human adaptation
and survival, we need to look to the origi-

nal, source categories which defined the very first inklings of survival in any mode or environment. By looking at the essential advantages, perhaps exceptions can be made to supplement those we have already adopted in order to secure greater comfort, and even to co-exist with other species. Some major exceptions occur which are worth noting, as a way of promoting survival in the abstract:

(A) Resource-dependence: unlimited resources could promote unlimited species.
(B) Niche: using different types of resources reduces the overall ecological footprint, and even produces desirable complexity.
(C) Efficiency: reducing resource consumption may result in greater survival, even with unlimited resources.
(D) Sufficiency: having a finite set of needs that must be accommodated works in tandem with efficiency to create highly paradigmatic Need-Demand thresholds.

Contradictory Deductions and Elaborate Conclusions - In reasoning about environments and organisms, it may be helpful to consider a proposition as though it could not be contradicted. However, it is clear that in some situations, perhaps all major situations that are investigated, conflicts emerge. Consider for example this rational refutation for a holocaust in an ideal society:

Dysfunctional Person Thinks: 'There should be a special place for me, this place should offer opportunities'

Sadistic Person Thinks: 'There should be a special place that disposesof them'

Dysfunctional Person Does Some Mind-Reading: 'Perhaps the place for me should-n't have opportunities'

Sadistic Person Does Some Mind-Reading: 'Perhaps there isn't a place for the dysfunctional person, or perhaps opportunities should be provided'

Government, Re-interpreting: 'Serving common interest, opportunities should be provided for everyone, or, at least, there is no sadistic place for a dysfunctional person'

Similar arguments might make conclusions as respectable as resolving relationship disputes by sex, or making governments democratic. It is likely that this method of compromise, or something slightly more complex, may be required to resolve disputes between multiple intelligent races. The simple form of the argument is *equity fatui*, that is, using value as a replacement for instinct--what the Greeks called 'the force of Eros.'

Cube of Life - A complex function of irreducible relationships;

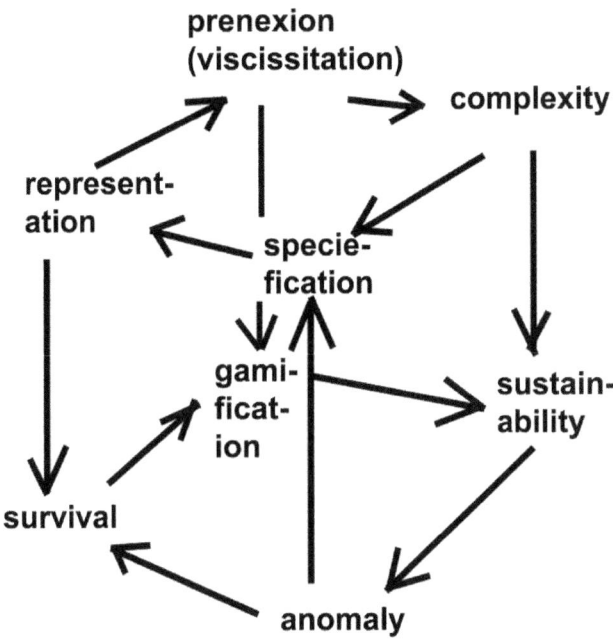

Here prenexion is a term referring to the emergence of new categories in biology, such as specific types of flesh, or other general adaptations. Anomaly refers to the organism as a whole. Complexity is the dynamic when it is considered that it can have a more-than-biological definition, in this sense referring at least to materialism or determinism. Gamification is the organism's tested performance as a species. The following offer opposite vantage points: representation and sustainability, complexity and survival, prenexion and anomaly, speciefication and gamification; The result is a perspective on life.

A categorical deduction of the diagram would read:

[1/1]: Complex survival is the vicissitude of anomaly when the representation of sustainability is the gamification of a species.

The role of sex, small necessary leaps, and social functions such as politics and personality can be seen here::

Culture - in the sense of the simplest biological makeup associated with an organism, it may be associated with aromatic or textural properties, such as the presence of coarse hairs, an earthy or fruity smell, or fine granules, seed-pods, etc., that rub off of the substance. It is frequently reported that members of a human couple are not offended by the smell of their beloved, so this could be considered a form of 'culture' which is subject to local adaption or coordinated chemistry. Similarly, the smell of lavender benefits not only by watering, but by the presence of elements in the soil which are more complex than merely earth or water. This may show a level of co-adaption between lichens and lavendula, which may be seen as a form of 'culture' similar to the interactions of simple plants in a petri dish. At the level of uni-cellular animals, culture must be considered to be predominately a case of symbiosis when it is not conflictive, but larger organisms can be considered as cultures of the interplay of many smaller bio-organisms.

Cyclotron - A simple idea that is becoming passé, is the general cycles and epicycles of existence, distinguished from atomic cyclotrons. What if occurrences happen in cycles, at specific times or through specific combinations of events? The inner biological cycle might then be ordered by exterior emergent events, which occur in cycles. There is then a choice to have a larger or smaller inner cycle than other cycle. For example, meals could be used for long-term planning, suggesting a long inner cycle, or meals could be used for planning the next meal, suggesting a short inner cycle. Features like these could be used to measure the comparative survival of different species, particularly by looking at the length of exterior cycles, and the distance between meals, copulations, respirations, etc. The overall result becomes a rubric for assessing *situationism*.

===========**[D]** ===========

Darwin, Charles - put forth one of the
most universal theories of biology, namely
the theory of evolution modified by a
mechanism of natural selection. Others
then applied the theory of evolution with
less success to spiritual, ontological, and
logical approaches. The theory of cosmo-
logical ontology had already been widely
refuted on the subject of theology (vis.
problem of evil) except in a very painful
valuistic sense, a sense which seemed to
lean on a theory of evil origins which was
not compatible with optimism or a divine
creation, and thus over-valued study of the
sciences as something supernatural for its
sheer grasp of anything---a truly ridiculous
standard, if people are not divine. Evolution set the stage for the radical post-
relativist stage where one form of life could
be superior, yet in which individuals may
still rise to the occasion by showing native
intelligence.

Decision (Metabolics or Meta-Biology) -
Early decisions in a plant or animal's life in
their pure form---in the sense in which development is purely self- or prior-
determined, may be as simple as the difference between 'learning' and 'leaning', or
between building and breathing. These

types of fundamental decisions are the types of decisions that affect consciousness. While teleology may take place through contractual schema such as fur in the winter or rewards for preparation, not all schema respond to a cause, or at least not an exterior one, and so therefore not all schema are contractual. This opens the door for some very exceptional cases, although they are also cases which must follow a rule of non-contractual development. While this is possible---say, by fiat, such as in the case of superb dinosaur vision affected by the coincidental crispness of plant life---there is some question as to whether mental properties such as volition require human brains. But since I take it for granted that some higher form may develop---at least a form more conscious than humans---it also seems evident that consciousness sometimes takes place in smaller or more primitive forms, although by the rule that it is exceptional relative to the level of nurturing that takes place. Perhaps exceptionism is the universal rule, suggesting that nurture plays a role which has nothing to do with the level of maturity of the individual, although clearly substantive differences could take place::

Deferential Exception - In highly specific cases of 'eternal' function or adaptation, it may be highly difficult to weed out details concerning every aspect of biotic function. This may be especially difficult in the case of micro-scale connexions over time and space. Functional contingencies may exist which resist the obvious or temporal normitization of definitions, and provides excesses and dearths of information. The highly specific nature of this potential excess and dearth may refer implicitly to correspondences, or explicitly, yet without explanation. What if, for example, lingering semen serve a function in the male urethra, an understated or informational variation of the female Barr Body? Or what if mushrooms are an altruistic species adapted to serve other species' mental functions? As potentially absurd as these theories sound, similar theories are potentially tenable in the complexity of time and space. The capacity for space-time correspondence is likely to raise contextual snarls in the ultimate realization of coherent definitions for species functions in future studies of biology::

Development of Growth -

theory	'serious-ness'
observ-ation	codific-ation

Diagrams read C.C.
from Upper Right

In the early life forms that are observed, codification has been adopted as a general form of life. Similarly, we can predict that human life occasionally succeeds to exist as a form of observation. If there were a more advanced race or condition, it might consist of the growth of theories, a form of 'supreity' (compare with virté under Uber-Adaption). A still more advanced form could embody seriousness itself. We know this instinctually, however, it seems rarely expressed. Each level has its own unique pattern of evolution, patterns which cohere, and depend on unique structural properties as much as viable dynamic interests. Not all of the forms have 'need' in the same sense: 'need, greed, name, read' seems to be the pattern running from codification to seriousness. A further theory is that the dependence on earlier and earlier forms of development in this diagram (which, I should say, are structures of every animal in the course of its growth), is a retroversion of the dependence on the hidden structure of physical laws in the universe; It can be predicted that if laws were not physical, early forms of development would be present at all levels, and the development towards an embodiment of forms of law would cease to matter. This is a rather grave statement on the predicament of evolution, particularly when it is hypothesized, as in my view, that the categories of developmental growth are exclusions of the only functional properties of matter; The alternate view, that biology is not the only form of life, is certainly radical!::

Dialectical Weight Problem - The general assumption that heavy people are surrounded by thin people, and vice versa, is highly disputable in the context of eating habits; What is more likely is a 'birds of a feather' mentality, where food or metabolism habits are shared in a group. Not only is it obvious that cognition may be mimetic, but it is also clear that any critical advantage or disadvantage needs certain influences to come into realization. In other words, even if a thin person amongst over-eaters doesn't over-eat, he is likely to change his standard of eating, and perhaps even his mental metabolic balance. Similarly, an 'indulger' amongst thin people imitates their habits, but more importantly, his or her own *relative* standard of eating and metabolism shifts. For this reason, the pattern of difference may be transparent and go unseen on survey results. The overall difference is one of context, including the wisdom (or lack of wisdom) of relatives and friends, and the sources of food available to the consumer. When food sources are said to be plentiful, then the search must turn to general patterns between knowledge and general food product trends. Clearly in this case we must say that 'significance is significant', even if it does not always spell cancer or boost IQ, it may be that the critical correspondence factor, based on group dynamics, is what makes the difference for health, statistically. Even if this factor cannot be detected in a general classification of individual food choices, quantification of mild factors and loyalty to major factors may spell the difference between decades of life.

Digestion - The best properties of digestion are what may be called efficient digestion or quick digestion, but which I prefer to call early digestion; Although the evidence is unconfirmed, there is some possibility that early digestion is more likely to occur in the upper torso than the lower; This carries the possibility of a hidden metabolism that follows no obvious rules: the rules are more inceptual or mental than they are enmattered. Early digestion may also be observed in the difference between alligators who make a meal and those that don't. Instead of ascribing their 'food mastery' to properties of anti-aging (although that thesis is clearly prevalent), I would like to hold that easy digestion is an independent property from lifespan, since lifespan and easy digestion vary widely from species to species and between individual organisms in most species. It is worth posturing that the best examples are more worthy of study, just as ultimately only the best medicine should be researched, accepting some relativity.

Dimensional Plants and Animals - In a basic view, dimensionalism is implied by any axial relationship, such as plant growth and the sun, or animal growth and the eating of plants,or a predatorial feast and the duration of the lives of the carnal animals. In a more advanced sense, however, dimensionalism only exists when it is present in the moment-to-moment existence of the species, a sense that is less asymmetric. For example, since plants are passive from a human position, they are likely to be perceived as having fewer dimensional properties; However, it is still possible that the 'passive' properties of a plant are higher-dimensional. For example, I have observed that reaching towards certain orb-shaped leaves produces the effect of extending the length of the hand and encountering the plant at a shorter distance.

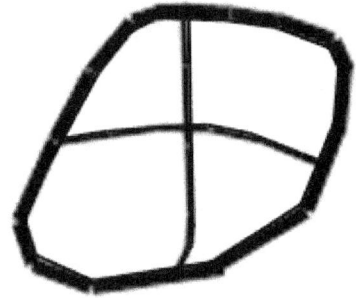

Typical higher dimensional leaves resemble a distended globe

Although this seems bizarre, it could be explained by a higher-dimensional curvature of the leaves. When these characteristics are not explained, that could be explained by mental reductionism, a form of ownership or consumption of information produced during an interaction with the plant. Cultivated species such as rhododendron may be especially prone to this reduction, even though they may be preferred for their beguiling complexity, such as an appeal to an aesthetic sense. The aesthetic effect is often divided between structural appearance and a singular character such as smell or color, reflecting in either case a propensity to be discriminated in terms of complexity: quantity or index.

Dimensional Animals: are distinguished by a special conditioning or 'narcissism' which determined that the character of the animal is to respond to specific or general environmental conditions, such as implementations or freedoms. One example of this is the small appearance of the hands of a highly dexterous rat. Although, actually, the hands are not very different from the size of other rats' hands; Thus, it is a dimensional property of adaption to higher complexity. Another examples is the independence of a giraffe. Much of its feeling must be in its neck compared to other animals. Thus, except in some mammal comparisons, the giraffe exhibits greater independence, communicated with the separate sensation of the neck. The same property of independent consciousness occurs in humans who are considered 'pariahs'. Although the sensation is not considered highly valuable, it does have perspectival advantages.

Most of the dimensional properties that have been observed related to scale; However, scale is not the only example. Through the presence of additional energy, or deployment of resources, small parts of the brain can make a paradigmatic difference. Such is the case with the adrenal gland in lab animals, and similarly in conventional survival and mating-related functions in wild animals. The role of this small difference in function should not be underestimated. Oftentimes a small biological difference is largely responsible for an organism's unique biological footprint. The small difference in genetic material between apes and humans is widely cited as evidence of the large impact of small differences in adaptation. From a dimensional point of view, these are 'dimensional differences'.

Now that dimensional differences have been defined, we can observe that dimensional characteristics are a large part of the identity of animals: (1) What an animal is capable of doing (defense mechanisms → tool use), (2) The ability to respond to a wide variety of situations (testing → general theorizing), (3) The ability to originate responses to nature (screeching → inventing), (4) The ability to uniquely take advantage of special opportunities (friendship → crowd-sourcing).

I argue that, while earlier concepts are definitive of some fraction of an idea, the functionality of an idea is ever-expanding. Likewise, the dimensionality of an organism expands through new modes and perceptions; It is largely the reduction or simplification of functions, often a function of limited contextual dimension, that is responsible for entropy at the level of animals. A psychological argument would say that the limit of context is the limit of thought. A biological argument on the other hand, holds that the generation of functional complexity is a response to specialized implementations, such as extremes of behavior and tools, or environment and neuro-biology.

Distributed Brains -

Distributed Brains like floating
brain clouds would generally
rely on a distributed network,
in which small damages by wind
or predators would remain
minimized.

Highly intelligent networks would need
to feed some sort of centralized member,
which would need to be inconspicuous
or else immune to damage

Dwelling Places - Some animals have the ability to create unique dwelling places. These generally involve the modification of the structure and function of the environment. The first extreme variable encountered, then, is usually a mere quantifier of the number of dwellings which have been created. Another factor is potential destruction of the environment which occurs during construction. If the new environment is to become global, it must present itself as a viable ecosystem, creating the problem of potential *environmental degredation.*

Organisms which build their own dwellings include humans, bees, ants, beavers, and many varieties of burrowing animals. There are also many animals, such as owls, woodpeckers, and clownfish, that make use of an already-existing feature of the environment with modification. Some animals, such as Titi birds, even dwell in the-mouths of crocodiles, cleaning their teeth! There are also extremophiles such as *heliophiles*, which are organisms evolved to live near the intense heat emerging from deep-sea vents.

================[E] ============

Enervation - plays a special role in many species, most notably the mimic octopus (*thaumoctopus mimicus*) which uses the shock of an attacker to beguile and intimidate; And jellyfish obviously use enervation as an offensive weapon, harnessing electricity. However, for most species there is a more passive role for this topic. Humans sometimes associate enervation with sex, much as mollusks enervate when they reproduce. If there is a clear contemporary analogy to the use of enervation, it is in the 'interface' of an organism, much as the nerves constitute the single best locutor of individual reality. E.g. frayed nerves transmit a frayed reality, likewise enervation embodies some action upon those nerves, not just an outward or inward action, but an action across boundaries. The concept of interface captures how enervation serves as a complex correspondence between the interior and exterior, as between the subject and context concepts::

Entic Property Development Levels -
    (1) *Life Level*: Beings develop in relation to a life-idea or life paradigm; In this state development is not psychologically social or individual, preventing researchers from understanding the exact mecha-

nism; In fact, this mechanic is 'life itself' independent of scientific assumptions or conventions of belief, qua techne;

(2) *Group Level*: In this type of context, individual development is obscured by social development, creating an illusion that all benefits are mutual, and derived from a real or fictional social construct or natural society meme; Here the larger idea of life-development has been obscured into a social concept, of any degree of consciousness, but some definite degree of organization. Hence it is at least proto-technical: it has a vocabulary or more crudely, a set of modes or 'moves'.

(3) *Individual Level*: Here development is attributed to the individual; Causes that are not desires may seem obscure. The explanation may be material, as -is, constructive, or ambitious. The social meme is interpreted as a function of quantities of individuals; The original life-force is obscured by the role of the individual.

Further complexity may be had by (1) Searching for deeper levels than the individual, notably by (A) Internalizing some principle related to life, society, or ambition and (B) Reflecting on the larger role this function plays for the individual, say technologically, linguistically, politically, or culturally, and (2) In some way extrapolating it beyond mere individual purposes, or by using a psychological theory of individual or social psychology.

The result should be neither conventional psychology nor pure individualism, but instead some mixture, with the effect that life or some meta-critical concept of society is incorporated to good effect. Besuiting biology, it will not be mere situationism, but a theory *of* situationism, nor can it be political without being macro-biological. If it is a group society theory, it should not depend on one detail, but should rather incorporate a general theory of individuals or life factors; In this way it may qualify as meta-psychology, and thus dimensional biology::

Epoché d' Ménàge - A term that may be used to describe the influence of synthetic or human-made paradigms upon animal existence; This is superseded by a number of things: (1) The humanness of animals and (2) The paradigmatics of animals and (3) The animalness of humans and (4) The synthetics of nature; Epoche d' Menage comes to represent a joking prevalent quality of the animal mentality; Essentially, its psychology. In the same way, when it irks us when it presents ugliness and brutality, or impresses us to adapt and alter itself, so too it impresses us when the animal melds and co-habitates as a pet or trophy sample; Although all of these examples are examples of adaptivity, they are also visible in coordination with humans.

It is as if the animal gains the quality of one of the superseding categories: The animal does one of four things: (1) Loses animal qualities, (2) Super-adapts, (3) Gains human qualities, or (4) Philosophically realizes the artificiality of nature. These things then indicate the representative psychological categories of animals whenever humans are involved; As such, the human comes to represent philosophy, while the animal comes to represent unrealized potential. It is a paradox::

Essentialist Biology - One biologist was quoted as saying that protoplasm is the 'endless flow' which is overtaking inorganic matter. Surely this is the essential nature of essentialist biology: to have one substance living and another dead, and for the living to make use of both organic and inorganic matter to perpetuate itself. However, an important criticism has been raised recently, that the inorganic is sometimes compatible with the organic. Some forms of inorganic compounds, such as nanotechnology, might have properties similar to those of organics. The question then merely remains, can a robot have a soul, a question which Wallach and Allen consider overstretched or supercilious. Perhaps a more definitive answer would be to say that for the time being, souls are alive, whereas robots are not, although the definition of a soul might change this. All of this has been assuming the rejection of an animist view in which love or some other force could provide souls for inorganic things. The religious argument is that any given thing with a soul was once non-living. According to this collection of views, robots can have souls in the same ways humans do, that is, if they have life.

Evasiveness - One of the most typical responses for animals in the wild is evasion. Although humans, being dominant, now often reserve evasion for specific scenarios labeled 'conflicts,' many animals in the wild continue to be subject to a wild pattern of continual fight-or-flight behaviors, often exaggerated during periods of mating. This is evident particularly in wild birds such as sparrows, and less so in more socialized animals such as geese and pigeons, e.g. because of the strategic advantage of numbers. In the natural ecological landscape human effects upon animals are often more pronounced in some forms than in others, particularly when the form of effect directly impacts the animals' natural habitat, such as food supplies or the ability to mate. Although today humans are accustomed to hearing that naturalized animals are only affected in drastic ways such as environmental damage, hunting, and climate change, in reality many effects are felt by animals (so far as this is possible) very early in the process via the instinct for evasion. The natural instinct to evade predators has the effect of creating dramatic differences in behavior

that are not directly or immediately correlated with shrinking numbers, and often
long before the most damaging effects take
place. Thus, some of the dangers posed on
animals may be attributed to maladaption,
when the root cause is actually the need to
evade predators or other perceived threats
(such as boats or hiking observers). Some
forms of animals such as woodchucks and
prairie dogs have already adapted to avoid
observation during long periods of their
lives---at least statistically. Others, such as
North American Deer and the Terrapin
turtle, continually run into conflict with
the order of urban society by impeding
traffic or becoming road kill. The influence
upon the ecosystem of Venice was probably felt very early simply by the use of
poles to guide gondolas (these could seem
like a threat to fish eggs which are stored
on the seafloor), although today a far more
disastrous although perhaps less axiomatic
effect is found in the common use of motor
boats. Even for humans, sound waves have
come to represent the threat of danger, as
exemplified by the whirling rotor blades in
an Indiana Jones movie, and the spooky
silhouette of a Hiroshima victim, whose
shadow is permanently marked on a concrete wall---schoolchildren inevitably associate this image with the sound '*boom*'.
This image is clearly analogous to the human predation upon wild ecological environments, where motorboats and rock music from hikers could seem like unknown
sources of danger. The animal response to

threats such as noise is often quick and ex-
hausting, and may perhaps be analogous
to human paranoia. Darwinism itself
stands on the assumption that natural
scrimmages occur in terms of survival.
Thus, there is no choice but to believe that
encountering conflict with other species is
capable of causing irrational and counter-
instinctual behaviors. Even if these behaviors are instinctual at their root, driven to
extreme ends even the animal performing
the behavior is forced to believe that it is
an exercise in futility. One of the few exceptions to the idea of defeatist statistical
biology is the concept of individuation in
which animals essentially get what they
want. In this view, the only exception to
the natural order is some kind of artificial
concept of authenticity imposed to promote realism to stand as an explanation
for otherwise vastly unnatural behaviors.
That is essentially the full account of evasiveness in nature, apart from individual
examples. For example, dimensionally, a
hypothetically conscious deer may find itself in captivity as it is found today,
whereas such a deer hundreds of years ago
may have believed it was a god of animals.
The idea is explained by entropy (the difference in energy level between otherwise
identical animals), as well as the prospective egotism of the deer.

Evolutionary Models -
Model 1: Survival Model. Species develop by conflict ('big issue') resolution .
Model 2: Reproductive Model. Species develop to fit body paradigm niches, such as combinations of wombs, stamens, ligaments, brains, etc. which influence reproductive success. This may include beauty, hormones, desirable features, etc.
Model 3: Adaptivity Model. Species develop to meet a holistic paradigm, such as symbiosis. This includes metabolic compatibility and adaptation to environment.
Model 4: Technology Model. Species develop by discovery of intelligent technologies (little-issue resolution).

Evolution in Four Dimensions - Godfrey-Smith considers this topic to be essentially un-simple and un-trivial. Evolution of this type might conceivably include features such as the following:

1. Dimensional characteristics, such as convolution, hidden abilities, and consciousness.
2. Use of space and time in a way that may be hyperbolic, especially immortality and time-travel.
3. Use of languages and other tools that involve the fourth dimension, for example, we have hints of this with holograms, quantum computing, and teleportation.
4. The four-dimensional or higher state of

125

the entire world.

Other authors (Jablonka and Lamb) use evolution in the fourth dimension as a title of a book, citing language and technology as fourth-dimensional factors. It is important, however, to qualify what type of technology or language is meant. After all, two- and three- dimensional language and technologies exist, and these are not necessarily fourth-dimensional. So what is meant by dimensional anything is necessarily either a generic qualification (dimensional typology), or else the nth dimension of the concerned properties.

Exaggerated Birds - The phoenix appears in some sense too exaggerated a figure, the condition of unpredictable fire and predictable re-birth; A similar doubt might appear in the singular dominion of the sky by birds and their imitators, the airplanes. Yet, where physical law permits, the dominion has some reality; Wings literally stretch as though to experience the entire sky, and the effect is one of dominion, however physically reduced. Indeed, people too seem to dominate, perhaps with less overall certainty, the physical domain in which they thrive; People have their arms free to be 'builders,' 'growers,' and 'nurturers,' and this is also a kind of inward grasping not dissimilar to the birds' outward sweep; What does it mean that people are inward, or are arms and legs more generic representations of wings or claws? Are we, in some sense, secrets of the earth, or are we imitating birds and projecting a project great distances, like something avant-garde? Maybe dimension is a plaything, or maybe we are functions of dimension, thus functions of our own minds; The question is raised, is there ia mind of nature, are there further thresholds to surpass? Are we replaced by our constructions? Ultimately: it is like the question of the phoenix: is it possible to combine chaotic energy and natural order? Is it primitive or sophisticated to spread our wings? Is there something wrong with imitating nature? The conflux of standards and paradigms may often rely on exemptions which are distinctly neither human nor anti-human.

Exceptional Biology -

9 Steps of Exceptionism in Biology

1. Survival
2. Leisure
3. Intelligence

4. Free cognition
5. Mental exceptions
6. Physical exceptions

7. Exceptional conditions
8. Exceptional dynamics
9. Exceptional existence.

Extra-Terrestrial Alien Brainstorm -

Here are rules to follow in forming an idea
of a functional alien life-form:
(1) Many similar structures may be present, with different physical functions.
(2) When it comes to reproduction, the solution is that 'nature finds a way'.
(3) Some aspect of biology is often unexpected, but this tends to be the easy part.
(4) Some obvious characteristics are defined by gravity or similar principles, combined with some possibly unknown defining characteristics, such as responses to
heat, or levels of conflict.
(5) The rule with absolutely alien species is
'same difference,' although this does not
deny extremophiles.

Examples of structures on extra-terrestrial
alien species:

Lip-brains
Heat-emoters
Finger-tractations
Organ-donors
Cognitive-stimuli
Memory-fetishes
Homing-sensors
Homing-defenses
Picture-armor
Complex-reactor
Mnemonic-mutation
Etc.

============[F] ============

Fallacies - *of Anthropology and Biology, Using A Coherent Method*

Diagram reads C.C. from upper right

[1] The Fault of Implementation: It is interpreted that a bottle rocket will send someone to the moon, however, this is false, a stronger rocket would work, but is designed differently. Similarly, some beetles are attracted to feces, while others mate by using clicking signals. It would be erroneous to think they were the same.
[2] Faulty Implication: It may be false to interpret that an officer fights in the infantry simply because he serves in the army.

Similarly, house cats are not necessarily dangerous to humans, even though they are descended from feral cats. This is typically a genus-species error, and it is important to biotechnology.

[3] Inappropriate Context: A very unstudious biologist may conclude that the male of a species is malformed when he is found amongst females of the same species. But he is wrong, the context has been mis-represented. Just because a sample differs does not automatically mean that it has no function; Indeed, only dysfunction would prove dysfunction.

[4] General Association Error: Let's say a man is assumed gay but it turns out that he simply has a case of flatulence. Again, an uneducated guess may stretch of truth of observable conditions. If one aspect continues to be observed, it may be that the prevalence is random, rather than explanatory. There must be some acceptance that statistical relevance is not always relevant to the species being considered, in terms of systems of meaning.

Family Resemblance—Genetic families are linked either by genetic links, or by genetic chains. A genetic link unifies the species by a single coupling or characteristic, whereas a chain unifies the family by exclusive or complete sets of characteristics.

The First Reverse Corollary - If evolution occurs towards some ultimate end---if there is purpose in evolution in a singular fashion, then this depends on creating just one type of offspring. It depends on combining multiple traits into one trait. So long as this isn't hypothetically possible, it seems satisfactory to determine that the future of evolution is multiplicitous. While there may be exceptions, they may also be dubious exceptions. The 'melting pot' mentality, for example, may favor long-term evolution at the expense of the short-term. Nor is replacement theory necessarily adequate either, under the view that major events are carefully constructed from trial and error. Replacement might result in a system that is overly fragile. If survival depends on short-term evolution, then the melting pot mentality does not serve. Although most scientists hold that evolution is strictly long-term, it may simply be that sufficient research or methodology has not been established for short-term evolution. It then seems that the correlation between short-term and long-term evolution is hypothetical at best. Although it may be seen that goal-oriented planning may lead to evolutionary shifts, some of this planning only makes sense collaboratively from an evolutionary standpoint. That is, it may be that the platform is hard-wired without the specific tools, meaning that although tools could easily be replaced, they would not always be present in the original (or most original) form. Therefore, it shows considerable optimism to say that short-term evolution is being preserved in long-term evolution. This may be the same as saying that long-term evolution is being replaced with the short-term, which is ridiculous. The reverse corollary is simply the statement that evolution occurs forwards through branching and variation, not through gradual decay of the majority. The exceptions to this, while possible (say, a through the incidence of the first-available universal environment*), are sometimes bleak. *The general trend of evolution, then, is something like the inversion of the least functional laws of nature.

Five Animals -

Systema, Utilizmus, Statika, Mobilius, Duribilis.

Systema includes any component of an organism that is especially highly organized. Because it essentially manages itself, it could be treated as a separate organism. Utilizmus is typically the human type, but could also involve adaptation to tools of other kinds, such as virtual reality. Statika is typical of plants, and mobilius is typical of animals. Duribilis is the immortal type, differentiated from any of the other types.

There are also a larger number of secondary types, such as Systema originalis (the beginnings of the branches of systema), Systema vulgaris (borrowed systems), and

Durabilis durabilis (meaning time itself as an organism).

Five New Categories Introduced with Biology -

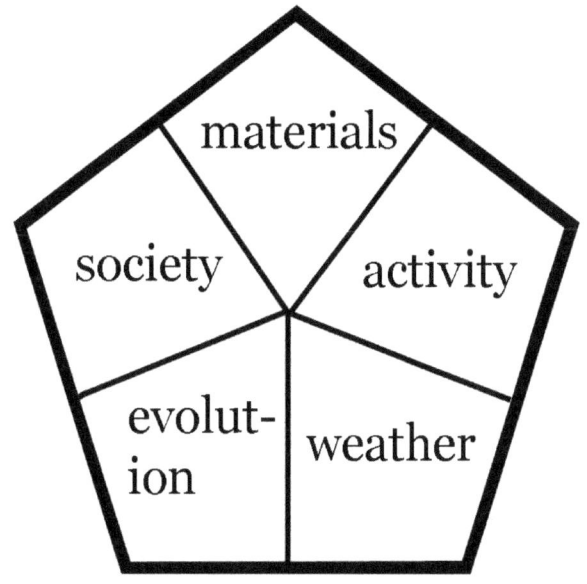

These also serve as the spatio-temporal referents for biology.

A further translation of the five categories, according to the penta-theory, results in the following statements:

[1] Activity involves a less material society, and a less weathered evolution;

[2] Materials involves less active weather and less social evolution;

[3] Society involves less material activity and less evolved weather;

[4] Evolution involves less social materials and less weathered activity;

[5] Weather involves less active materials and less evolved society;

On the other hand, the following statements could also be made:

[A] Activity is social evolution
[B] Materials are evolved weather
[C] Society is active weather
[D] Evolution is the activity of materials
[E] Weather is the material society

Although, depending on perspective, it may appear that each of these items occur only at some specific advanced stage in history, actually the patterns are recurrent, and can emerge both in little and in large ways.

Foods - *Palatable Foods*

Produce, Produce, Produce, Produce
(Spiced, Fresh, Manufactured, Product):

This sums up the foods that people eat.

Produce (fresh)	Produce (spiced)
Produce (manuf-actured)	Produce (product)

Diagram reads counter-clockwise
from upper right.

It sounds scatological, but there you have
it. Most, if not all, foods that people eat fit
into one of these four categories. That is
not to say that the categories automatically
qualify any given thing as food. To the contrary, what is expressed is a minimum
standard qualification for consumables.
Combining the categories is even more apt,
with the correct understanding, to yield
palatable foods. Additional insight may be
had by considering climate and biological
needs, e.g. there may be an attraction to
cold foods in hot weather, warm foods in
cold weather, and dairy foods in general.
Alternately, a complexity model may favor
meat-based foods, a mass-consumption

model may favor grains or other products that can be mass-produced, and in general there may be a preference for food that is easy to get, and which is not rotten, fitting the second category ('fresh'). These four models and their additional interpretations may be one of the most important formal constructs for understanding the nature of palatable food, as well as the need for a varied diet and good nutrition. (1) Available (2) Easy (3) Substantial (4) Palatable.

Four Plateaus of Civilization -
Lowest: Translation-Level: brute feelings like pain and pleasure.
2nd Lowest: Nominal Suffering: it is possible to acquire taste and preferences.
2nd Highest: Idea of Life: through knowledge, the desirable things are understood.
Highest: Idea of Reality: through intelligence, the greatest ideas (the idea of reality) become part of life.

'Foyering' Vs. Genetic Assumption -
Science makes room for two views of progenitation; However, only one has been accepted; Genetic assumption is the type in which genetic data 'dominates' the mating process. If we assume that some choices realize the data native in experience, or must respond to the limit of prior modalities, it is also possible there is another type, in which the organism dominates the genetic structure. Although to

some minds this suggests or evokes supernatural control over given attributes, the
fact that those attributes are given at birth
may put a limit on supernatural appear-
ances---E.g. I do not usually mean a sense
of outward mutation, like turtles with two
heads. The second type is called 'foyering',
a way of placing genetic code ahead of, but
supervenient upon, other data. One form
of this, however, is merely a more complex, yet less obvious interpretation of biological data, much as humans are sometimes but debatably, set far apart from
chimpanzees. The argument is not that
there is no genetic difference between
these second types of humans and the first
type, but rather that the genetic expression
may be less obvious than in the first type.
The absence of an obvious difference from
other humans is explained by the need to
mate. Thus, the 'foyered' type may differ in
its mating habits, while remaining outwardly the same otherwise. Thus one may
characterize one of the following categories
to express an individual: (1) Genetic assumption dominates real mating situations, so foyering tends to be recessive, or
(2) Foyering is a dominant trait of selection, so genetic assumption is recessively
dominant---useful but not centrally useful.

Fractal Biology -

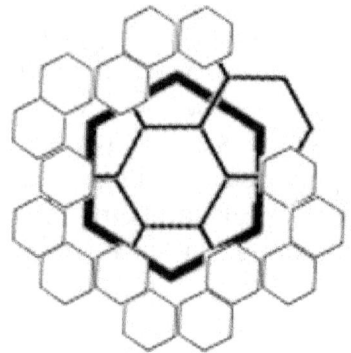

Exists in distinct forms in ordinary physics:
1. The relationship of a structure to a crys-
tal lattice, e.g. planck lengths of a black
hole; 2. Relationships by causal organization; 3. Relationships by conceptual organization;

Thus, there is an implied hierarchy of (1)
Energy, (2) Possibility, and (3) Imagination, implying that in non-arbitrary cases,
fractal biology is either energy or imagination, but in all arbitrary cases it consists of
pure possibility or probability; Thus, there
is some choice between systems, but
imagination is the most open-ended,
whereas energy structures is the most
highly studied;

If dimensional biology is a step beyond dimensional psychology, then in all cases
imagination is subject to some kind of

structuring; The implication then is that the two types are directly correlated via probability; Then it is determined that fractal biology is indeterminate only where there is volition (undetermined energy) or sheerly probabilistic matter, in the form of energy that consists only of probability; Therefore, where energy reduces to something other than probability, it also reduces to volition or determinism; Although this is contradictory, it explains how fractal biology is a contingent phenomenon, a perspective on concepts, probabilities, or energies which may otherwise be otherwise organized.

Fresh Air and Sunlight - Water, fresh air, and sunlight has been advocated for potted plants as well as ailing intellectuals with dysentery. There have been wide remarks made about the value of the fresh, balmy air that preceded the industrial revolutions. Perhaps it is even a vital thread to the common properties of animals, or a vital link to the previous, most enjoyable experiences of ancestors. The 'spots of light' or 'spots of time' theory might go far towards explaining the virtues of pleasurable experiences for genetics. Clearly this is not to say that these experiences are automatically impracticable or impractical. Instead, practical experiences tend to be those which are best remembered, or at least, which trigger the best long-term responses for an organism. These experiences need not be those ex-

periences which seem *rationally* to be the
most practical. Instead, they tend to be experiences which appeal to conservative
survival, including combinations of such
terms as mindfulness and water availability or laziness and patience, for example.
These types of correspondences can easily
be programmed through successful actions
in the theatre of life. Fresh air and sunlight
(or any other successful resource paradigm
for xenoid organisms, when it is viable),
becomes an even more important level of
survival, a level which it may be said, has
less to do with individual organisms in the
long term than it has to do with ecology
and perhaps social organization (e.g. because social organization affects survival).
Thus, there are three primary levels of the
fresh air and sunlight paradigm: (1) Natural Resources, (2) Social Organization or
Supreme Adaptability, and (3) Individual
Co-Adaption and Consumption. At the
third level, it can be seen that there is a
strong role for provider and beneficiary
behaviors, often in the form of a specific
adaptation, such as leadership in animals
or fillistary roots in plants.

Function Taxa for Plants and Animals -

A SKETCH OF FUNDAMENTAL TAXA OF FUNCTIONS FOR PLANTS AND ANIMALS

[TYPOLOGICAL DIMENSION]

1: 0-POINT

SCRIMMAGE OF INVERSION:
Secret Key: base function and dimensionism / complexity

2: LINEAR

A. Scrimmage of Reaction
Secret Key: Infinite Reaction

B. Scrimmage of Process
Secret Key: Permanent Process

3: GRAVITY

A. Scrimmage of Confrontation
Secret Key: Quantity Vs. Scale (Property)

B. Scrimmage of Application
Secret Key: Allocation Energy (Form)

C. Scrimmage of Adaptation
Secret Key: Idea >= Recursion

4: TIME

A. Scrimmage of Growth
Secret Key: Add (Block) and Subtract (Motion)

B. Scrimmage of Impacts
Secret Key: Defense (Offense) and Communication (Interpretation)

C. Scrimmage of Swallowing and Disgorging
Secret Key: If-Eats (If-Edible)

D. Scrimmage of Surface
Secret Key: Attraction and Repulsion

See also the corresponding diagram at Meta Functions.

============**[G]** ===========

Game State [Diagram] -

Example:

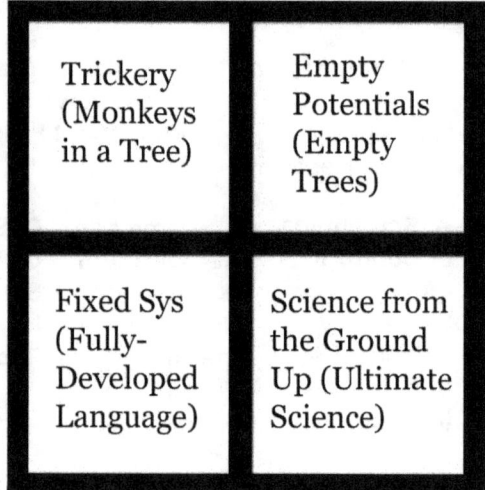

Trickery (Monkeys in a Tree)	Empty Potentials (Empty Trees)
Fixed Sys (Fully-Developed Language)	Science from the Ground Up (Ultimate Science)

Categorical Deduction:
(1) There is an empty potential for trickery when a fixed system is the ultimate science.
(2) There is an empty potential for the ultimate science when the fixed system is trickery.

Conclusion: Science exists in multiple degrees which correspond roughly with language development.

Gay Dilemma: What is sometimes called gay in nature sciences is excessive exhibitionism (exhibitionism to the point of weakness). In the human species, this type of behavior is associated with a lack of aggression and low tolerance for pain. These are characteristics which some societies eschew in a mate, and which other societies merely disregard, in favor of types of signs of machismo, such as economic success, intelligence, or social functionalism. If these second characteristics are the new replacing the old, then apparently evolution has found no alternative to entropy except divinity.

Gender [human]-

History : Social dynamics of personality/
social evaluation
Perfection: Social typing or Self-typing
Complexity: Asexual attraction

REVERSE:

1: Pure Attraction
2: Original   or   Social
3: Dynamic,   Personality,   Evaluation

This reveals several things: a highly polar conditioning of gender, a dependence on secondary but not tertiary types except under conditioning, and a type system that is not explicit; This explains phenomena such as the following:

[1] The strong role of personality
[2] The dubious but potent role of polarized factors such as superiority and anger
[3] The change of sexual attractors over time and through history

As a function of those declarations, there are further insights, such as the contingency of specific conditioned responses to sex, the explicit role of mental capacities in defining ambiguous functions, and the high dependence on social conditioning, or any creative factor that manages to replace it. See also under Appendages (Sexual).

Genetic Deduction - *Coherent genetics using simplification.*

Opposite boxes are treated as opposite categories. Categories are arranged in coherent sets of four, as shown (a scheme like this can be arranged for any size set, as proven by symmetry and dimensionality):

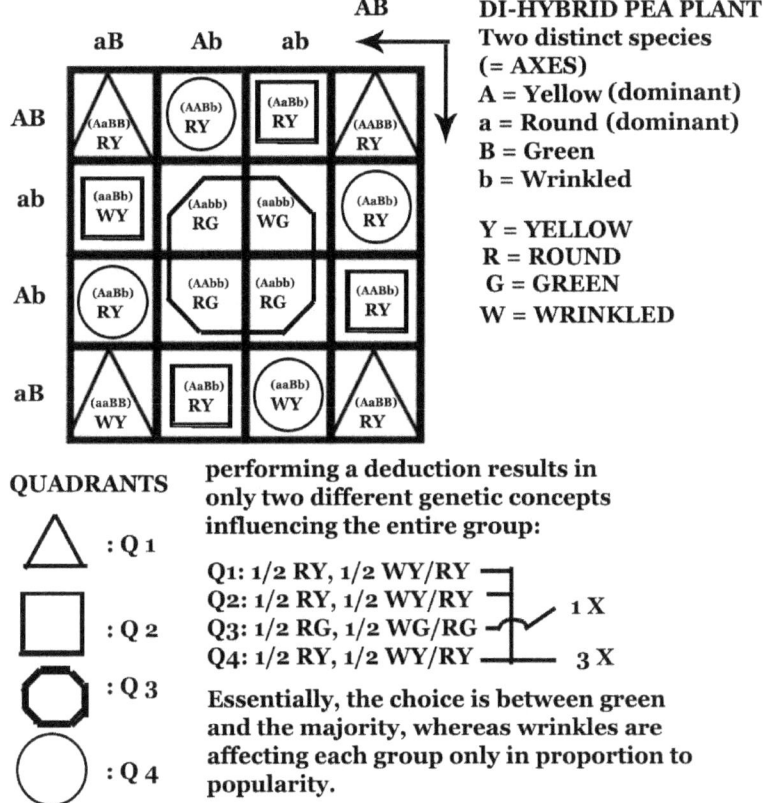

DI-HYBRID PEA PLANT
Two distinct species
(= AXES)
A = Yellow (dominant)
a = Round (dominant)
B = Green
b = Wrinkled

Y = YELLOW
R = ROUND
G = GREEN
W = WRINKLED

QUADRANTS

△ : Q 1

☐ : Q 2

⬡ : Q 3

◯ : Q 4

performing a deduction results in
only two different genetic concepts
influencing the entire group:

Q1: 1/2 RY, 1/2 WY/RY
Q2: 1/2 RY, 1/2 WY/RY   1 X
Q3: 1/2 RG, 1/2 WG/RG
Q4: 1/2 RY, 1/2 WY/RY   3 X

Essentially, the choice is between green
and the majority, whereas wrinkles are
affecting each group only in proportion to
popularity.

Each four-part quadra, representing a coherent set and
corresponding to the different shaped boxes, are counted
and sub-divided into two-category sections. Each two-part
set of two is analogous to a categorical deduction of the
form (A-B:C-D and A-D:C-B). In biology, however, the
logical method is not necessary. Instead, categories are
grouped according to whether they share identity with the
other groups. (If they do not, this means essentially that
dominance was irrelevant, because the organisms had dif-

145

ferent genomes entirely. In that case, it is
unlikely that they would reproduce). The
final result gives a ratio of influence that is
far simpler than the original assessment,
as shown at the bottom of the diagram
(one has a 3/4ths ratio, one has a 1/4th ratio). Because of the common characteristics of hybrids, and because this combination originates with the most diverse possible configuration, it may be trusted that
some of this simplification appears in the
form of a general rule. In other cases, such
as first-generation crosses, the result will
be a different ratio of simplification, such
as 1:2 instead of 1:3. The simplest cases are
not open to the same degree of simplification. The result shows influences upon the
individual examples in sheer statistical
terms: the simplified reality is that 3/4ths
are influenced by a combination of round /
yellow (¾) and wrinkled yellow (¼), and
¼ is influenced by ¾ round green, and ¼
wrinkled green. Thus, there is really a gra-
dation between common characteristics
(yellow) and permitted flaws (wrinkled
green). From this general characterization
of hybrids, we can reach for general laws of
genetics:

Laws of Genetics

(1) With dominance, success becomes predictable.
(2) If a flaw dominates, desirable
characteristics become the exception.
(3) The exception within the norm

dominates the exception within the exceptional, unless the purely exceptional is desirable, and the desirable is successful.

(4) Generally, the organism is only as functional as the norm, except in terms of exceptions.

Genetics -

N-Modifier: As the chance of genetic modification and mutation increase, there is an increasing likelihood that examples of original species will also be nominally related to the actual original; For example, plant species may exhibit patterns of adaptation to technology or environment (such as radio waves or radiation) that go unobserved for generations; Consequently and increasingly, the viability of the authenticity of a species outside of an adaptive thesis is equal to a ratio to N, where N represents the authenticity of the name of a species; Clearly, the name of a species may become obscure, as medical or corporate conventions are adopted; There is also an increasing risk that examples of primary species will have inter-bred with a modified species or become altered by the environments; The risk here is idealism, but also apparent functionalism; Thus the shift seems inevitable; The N-modifier has an earlier precedent where N- equals Nature, so that all things are relatively themselves, unless the world is synthetic; This has complex interpretations when N is not 1 or

0, such as when adaptation is radically contextual; This secondary thesis can be used to correlate with the first.

Genetic Semantics (Rhetorical Biology) -

1. By the time genetics is fully understood, we will be well beyond it.

2. Or, maybe genetics is a sinkhole in space & time.

3. Genetics must record both past and future developments (sacrifice).

4. Or, genetics may be termed non-metaphysical.

In four, genetics involves a context for being other than a recording, which duplicates rather than genuinely learning. One thesis lemma may be that learning only takes place through cognitive function, if genetics is merely a recorder, e.g.

5. There may be a differentiation between cognitive and non-cognitive genetics, that mirrors states of development.

Genetic Weirdness - First I tell myself that my mind should be weird, and my body NOT weird. Then I say my children are not to be weird. But, unfortunately, if they can't consider weird, and moreover if their parents' minds are weird, there's no resistance to being weird! I come to the conclusion finally, that weird means MORE WEIRD. Normal people of any variety are weirdly normal. For example, they might be highly intelligent, and thus display a normal medium of entertainment. Or they might be weirdly habitual and predictable, thus out-performing everyone else at 'ordinary' tasks (which don't seem so ordinary, because no one else can do them all the time). Or someone may have a vivacious personality, thus demonstrating what everyone else would like to live up to, but is unable. In any of these cases, it may be difficult to recognize how exceptional the personality really is. The assumption is that they are over-performers, when actually, everyone else is under-performing. How does this apply genetically? Well, replace weirdness with the dominant gene. Normality then translates as some sort of dominance, or else weakness. Normality makes it easier to decide contested mates. If the male's dominance is in question, we can ask, is he normal? If he is normal, we can ask, is he dominant compared to everyone? The female can then ask, is this the dominant gene that I prefer? At that point, everything can be settled by conflict or polygamy---polygamy usually meaning dominance by one or the other sex.

So, working backwards, there is dominance or conflict stemming from female

fetishes, concerned with universal dominance determined by normality or unques-
tioned dominance, the root of which is the
truth that what is actually dominant is
functional weirdness, in other words, the
character of dominant genes. However,
again, to express weirdness as a mental
trait, or to have a body that is more weird
than a normal concept of weird, leads to
trouble in developing children. So, essentially, secondary traits consist of reassurance, not about the nature of weird, but
about the nature of reality. This is the conventional parent trap, where children become alienated from the reality of the parents' partnership---such as by the hyper-
bolizing of its meaning, or the genericizing,
the de-characterizing, of its role or functions. Both of these difficulties lead to a
characterless idea of parenting, a need to
fulfill a superficial alter-ego. Psychologically, this is just the presence of secondary
partners, justified by lust, and the decep-
tion of dominance. Thus, some of the
deadliest psychological aspects are concerned with false dominance, and ways in
which desire is corrupting. It is easy to see
how the image of a bird, once innocent,
can begin to represent the characterless
activity of lust or wrath---all bent on the
aims of more powerful selves, and endlessly sacrificing. Sometimes the central
gift of genetics is escaped under the mere
conditionalism of parenthood.

One conclusion is that much of it is a performance, more superficial than the mind
can recognize. Another angle is that what
is most important is not the larger picture
of genetic dominance, but rather, the writ-
large features of everyday experience,
things which are themselves exemptions---
at least, a kind of exemption---from the
world's history of conflict and pitfalls. The
conscience of the parent sometimes re-
bounds, recognizing that in children, are
embodied many, perhaps worse sufferings
than the parent themselves endured in the
process of maturation. For this reason,
events like graduations, birthdays, and re-
unions acquire a kind of pathetic piqué.
Perhaps the self-consciousness of the out-
liers in these events is a self-consciousness
about the nature of dominant traits?

Geometric Variations - in plants and organs

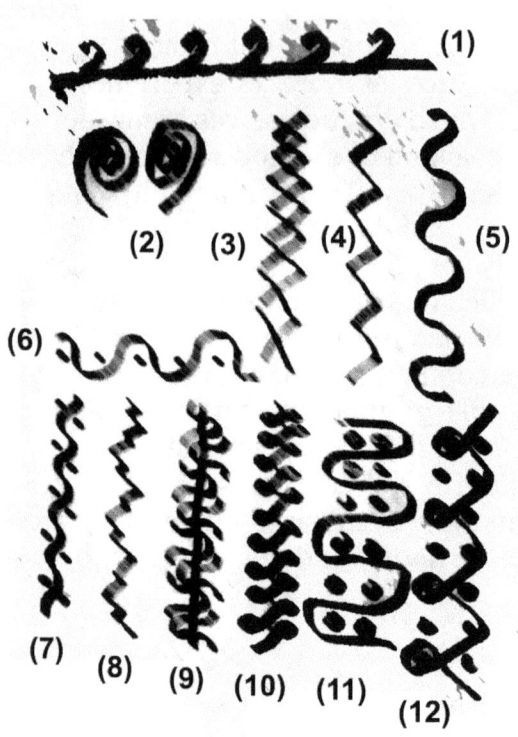

1. Simple growth, 2. Coordinated growth, 3. Simple organization, 4. Simplified organization, 5. Efficient organization, 6. Organized coordination, 7. Efficient coordinated organization, 8. Additional dimension, 9. Organized complexity, 10. Efficient complexity, 11. Efficient organized complexity, 12. Additional complexity and organization.

Gestalt Biology - Let's consider some statements:

The body is the brain: Reductively, mimetically, or pragmatically, or metaphysically-then-physically - But if the body is the brain, it might also be part of the brain, leading to an infinite regress. Apparently, to believe in the mind is to believe in the soul.

And, there is a poem:

Nature turned to yellow
When Frost thought of nature
The tree of all knowledges
Grew out of divinity
The egg in the branches
Turned with infinity
The egg grew inside
The contours of space and time
What blossomed: a mortal mind
That thought it true
What it had divined

See also the list of organisms in the appendix. Some of these are based on a gestalt theory.

============**[H]**============

Hennig, Willi - The founder of the biological science school of cladistics, which assesses species from the orientation of the overall tree of life. Diagrams can be formed which show species, often branching from one into many. The branches can be measured by the size of the population, or the importance of the organism's genes relative to the others.

Higher-Conditional Reality - Higher-dimensional conditions for reality do much to explain the abstract relationship between contingent existence and ersatz manifestation. However, arguments in favor of it have been scarce, whereas opposition has been numerous. Arguments have been proposed, perhaps half-unconsciously against the dimensional perspective, favoring the view that limbs and opposable thumbs emerge *ex simpliciter*, because of gradual improvements, which is a strong thesis until it is realized that the potential reality must be much more complex before even the simplest changes are allowed to take place. I have no opposition for the related argument that evolution is a function of time. However, consider the hypothesis that shortcuts may be made in evolution through modes of perception or learned behavior. I grant that these may not be strictly 'dimensional' loopholes in reality. They may be more conventional than super-natural or quantum. But quantum reality has already been accepted to exist on a sub-atomic level, so I see no reason not be believe that it could entangle itself with so complex an organ as a *brain*. Now, if it is understood that there is some degree of built-in relativism, namely that some events that may seem highly significant to an organism are later deemed irrelevant to evolution, or are even unobservable---then the role of the brain in relation to evolution becomes more subtle, and also, I think, more obvious. It seems clear that evolution is not inherently the manner of comprehending the changes that take place within the brain. So, if we ignore evolution, it is much easier to ascertain that some change might take place which could be called dimensional. After all, evolution is a sort of absolutist frame of mind, which is somewhat demanding on the subject of what constitutes a dimension. The human body form with all its precedents, and the functioning brain are sometimes counted as the only major accomplishments of evolution. It is therefore important, from a dimensional perspective, to consider evolution as multi-variant or else to consider that it is not the primary, or at least not the initial, mode for comprehending life forms. Other modes, such as psychology and physics, might be better adapted than evolution in certain forms. I hate to say it, but this again calls for multi-disciplinary approaches, which in turn call for a more ascertaining abstract method. I have described the beginning of what I believe to be such a method, which may be frivolous, in my writings on *Metemphysics*. The discipline is concerned with the materialism of ideas.

Higher Functioning, 4 Levels of
To be compared with other systems.

Level 1:

Strength, Speed, Intelligence	Relative advantageousness, Long-term survival
Weak bladder, Paranoia, etc.	Wealth, Leadership, Originality, etc. (Paradigms)

A. Measure of weakness.
B. Measure of strength.
C. Detractions / compromises.
D. Paradigmatic functions

Level 2:

A. Health., B. Happiness.
C. Giving., D. Authenticity.

Level 3:

A. Superiority., B. Knowing.
C. Accomplishment., D. Emotion.

Level 4:

A. Preservation.
B. Value.
C. Virtue.
D. Control.

Hollow Biology Problem - Sometimes philosophers of biology confront what appears to be the onset of nihilism. Somewhere between the Mandelian Law of selectivity, and the insistence of individual genes is a puzzle which only expands when genes have complex expression. How much of genetics is a factor of environment? What constitutes fully determined reproduction? One answer is by-the-textbook, and says that organisms follow prescribed patterns. But in recent years there has been a growing urge to understand, just how much of this is probability? What is the real role of chemical 'tags' which accompany RNA and DNA? How much is pure transcription, and how much is something more complex? One resolution to this difficulty is to accept that there are varying degrees of qualification for biological circumstances. The difference between someone who cares about the specific, versus caring about the general, versus people that do not care in specific and general ways, has not yet been mapped onto the system of genetics. Individual variation is seen as being too great. And it would not do to simply accept determinism in place of volitional brains. But if the nihilistic problem is to be overcome, general rules must be found, patterns must emerge, to explain the chasm between functional and complex data, and mere textbook cases.

============**[I]** ==========

Immortality - A number of factors may contribute to biological immortality, from the standpoint of biological knowledge and survival. The obvious factors include natural immunity, physical endurance, and intelligence on matters of survival. Other factors may weigh in as well. Some of these factors weigh in later in life, such as the openness for physical mutations which would extend life ability, counteracted obviously by disease mutations which pose a danger to health. Some people still consider natural mutation to be a sign of demons or magic, while others simply find mutation difficult or undesirable. Nonetheless, mutation, while not often considered practical, has some categorical usefulness. Experts on aging, for example, have often lauded the human ability to remain mentally alert and respond well to stress later in life, and these abilities are in some ways similar to mutations, if a mutation is compared to chemical patterns which engage at a certain time, while remaining unstressed at others. Perhaps a better way to summarize the willingness to mutate might be the body's metabolism, and ability to process energy. Someone who readily processes energy might easily feel more alert and quick-witted. So, one of the major factors in determining longevity in the context of biology might be the body's adaptive chronology, resulting in nine different patterns:

1. High energy, efficient output
2. High energy, specialized output
3. High energy, high output
4. Efficient energy, distributed output
5. Efficient energy, normal output
6. Efficient energy, low output
7. Low energy, normal output
8. Low energy, low output
9. Low energy, no output

In general, immortality seems to divide into several types, including metabolic immortality, cognitive immortality, and cellular immortality.

Metabolic immortality is the type most likely to occur through adaptation, but sometimes may require the shedding of dead cells. Cognitive immortality allows information to survive, but may require life support systems, which can be economically dependent. Cellular immortality is the type of immortality envisioned by medicine, which would guarantee that cells can live and reproduce, but may be prone to genetic aberrations.

Incomplete Information - Although fictional arcologies can be devised showing any sort of hypothetical survival-habitat relationship, there is some question in the real cases about whether all information can be known on the subject. Certainly, just as in the fictional cases, information can be formalized to some degree, to express predictable outcomes, but this does not always explain the true and genuine phenomenological aspects of a species. For one thing, new and emerging species do happen, and in some ways these new species are inevitably on the forefront of what it means to be adapted. If old adaptations depend on a certain future flexibility, then patterns can be mapped in terms of old adaptations. But if it turns out that old adaptations do not explain new activity, then an entirely new method must be devised, that accounts of new forms of adaptation while not losing sight of the evolutionary scale of biology. If it turns out that there really is a 'local singularity' of increasing survival pressures, or sudden environmental shifts which have occurred only recently (e.g. due to humans), then it behooves us to explain the broad evolutionary change in those terms. There is a necessity not to fictionalize biology, but there is also a necessity to appreciate biology on the terms that it exists today (or at any time in its evolutionary history).

Individual Pluralism - one of the understated sources of genuine government, individual pluralism is the presence of multiple concepts of individualism within the social organism; For example, one may correspond to public roles, and one to private, or there may be different roles for every type of ritual function; Roles are especially influenced by concepts of reality, which may become segregated according to obvious sensory differences, such as time or place; The difference between functions---and realities---becomes a basis for comparing determinative actions to decisions based on ostensive correspondence to de-coherence between definitions of individualism; For its opposite, see Social Argocy.

Ingestion - There are primarily three forms of ingestion amongst animals; In the case of single-celled life-forms, it may be expressed as exaggerations of these, in a sense that is more coincidental; The forms are neutral ingestion, positive ingestion, and negative ingestion; Descriptions follow:

Neutral Ingestion - is the property of processing or partially absorbing moisture or energy from a source that has not been admitted into the body; This source could be suspended in many stamens, fronds, or ganglia, or instead squeezed by the outer surface of the ingesting plant or animal; A primary example is the case of herbaceous moss, which seems to gain nutrition from leaves and perhaps even insects, without destroying or ingesting their entire mass; Oftentimes this process is entirely harmless to the food source; The use of neutral ingestion for example on a skin surface becomes an interesting future prospect for biotechnology and lab-created organisms, in the context of the high energy food sources such as heat and radioactivity; This is in keeping with the electronic paradigm in biotechnology, which has the potential to develop even further with 'gray goo', genetically-altered organisms, and the specific highly chemical responses between energy states and genetics or other codified processes, analogous to parallel and serial processing;

Positive Ingestion is full absorption of a food source into the body, either by repetitive chewing or sucking, or some sort of complete gorging of an edible body; This often requires musculature or an acidic enzyme, and may constitute a large part of an animal's energy commitment; The large commitment opens the prospect of herbivorous dependence on low-energy goods in bulk, or more isolated gorging on selected high-energy foods; Even when there is not a time commitment to hunting or gathering food sources, the category of processing may affect activity or learning ability; Other developments, whether or not they are related to the dynamic with the food source, may depend on the commitment to the high or low energy paradigm, or secondary characteristics which have evolved prior to the food adaptivity behavior; This is less the case with neutral ingestion, which is essentially an effective or ineffective application which may be extraneous to reproduction; And unlike negative ingestion, this paradigm is less reliant on the adapted patterns of other animals or plants; In this sense, it is a less specific and more general response to life conditions;

Negative Ingestion - exists for passive predators such as spiders and jellyfish, and parasitic animals such as a hookworm or maggot, and conceptually, even a vulture; In these cases the avenue to a meal is highly correspondent with reproduction, or in some way determines the nature of social behavior; Usually there is a subtlety with the boundary between predator and prey which renders the exterior dynamic dangerous for larger or smaller animals or plants; Unlike in the case of positive ingestion (which is more moderate), the difference between the size of the predator and the size of the prey may be highly determinate of the dynamic of the relationship; Although negative ingestion entails a certain amount of adaptation, this adaptation may correspond to environmental dependencies which potentially have only a contingent form in genetics; If it is argued that that is the case with all genetic adaptations, it can be stated at least that positive ingestion is more moderate in its preferred climates and conditions; When the climate is extreme for negative forms of ingestion, it may be a function of climate adaptation rather than the specifics of the victim.

Instincts -

The Four Instincts
The impulse for nature
The impulse for society
The impulse against society
The impulse against nature

Four More Advanced Instincts:
The impulse for a new nature
The impulse for an old society
The instinct for a new society
The impulse for an old nature

Interesting Species of Animals-

Common hummingbird - remarkable wing speed.
Birds of paradise - a real species.
Whale shark - the largest mammal on earth
Queen ant - different species from the other ants.
Komodo dragon - the largest lizard on earth, after the fall of the dinosaurs.
Phyllium Giganteum - Walking Leaf Insect
Thaumoctopus mimicus - Mimic octopus
Botryllus schlosseri - Golden Starfish - 77% human DNA
Phromina: Parasitoid that inspired the *Alien* movie.
Mola mola - large fish with no tailfin
Liparidae (snailfish) --- deepest fish ever found: 4+ miles down..
Homo sapiens sapiens - the most philoso-phical sub-species.

Isomorphic Breathing - like osmotic breathing, is an efficient form of oxygen absorption; In this case it is breathing from or into a distant or unknown (or unusual) source; Such possibilities are broached by the existence of oxygen teleportation technologies, or perhaps natural quantum or heat properties, e.g. such breathing may also be called quantum breathing, and consists only of useful information (or larger, information patterns), and not literally oxygen; This has the prospect of increasing metabolism of ordinary oxygen traces, as may be reported with exceptional athletes and sometimes under the surface of blankets; Isomorphy may be combined with low oxygen usage, e.g. the replacement of oxygen with some other fuel source through an unknown process; Perhaps the blanket theory is explained by air compression or adhesion, but this does not explain both the longevity of its usefulness and the method of efficient intake; One associated theory is the use of other gasses such as carbon dioxide or methane for breathing, as in some sort of fungal process, but such theories are often ruled out by common biological viewpoints, returning over and over again to the statement 'available oxygen', which may in turn be questioned.

Nathan Coppedge

============ **[K]** ============

Knowledge Infestation - A useful concept on the boundary between humanism and biology is the concept that some forms of knowledge seem to have a life of their own. Students may find that moving fingers like windshield wipers is a good dish cleaning method, or that large horse flies are sometimes attracted to lemonade, because it is known to kill other types of insects. The patterns that emerge connect techniques to bio-systems, often by triangulating an object, a subject, and a location. There is often a trigger, such as a classification of location, an action on the part of the subject, or a requirement for the object. In this way, student learning functions much in the same way as the acts of nature. The question for the scientist is, how to change behavior without contradicting instinct? That might require an epiphany.

========**[L]**===========

Lamarckian Twist - *Named after August Lamarck*. The impetus to survival exemplified by intelligence specifically, especially I.Q. or genius and it's correlation with reproduction. Some consider this to be an out-on-a-limb conjecture, whereas others consider it to be a self-solving point, which it could be useless to refute. Although perhaps there are things more survival-worthy than intelligence (one can imagine some sort of pleasurable virtue, or magical dominance), the point stands according to J.S. Mill that intellectual phenomena are the most deserving of pleasure, and any higher virtue must thus be achieved through them. What this says about sexual pleasure is rather mysterious, as it seems that although sex would be greater in intensity, it could be subjugated to a lower tier of value (say, ethically). However, the biologist has no choice but to consider sex as something at least equal in importance to intelligence, unless life consists primarily of the experiences of trickery to that end. It would then be a truly devious world, in which all value could be corrupted by the ability to choose mates arbitrarily. In this type of world a Lamarckian might say that the virtue---which may or may not seem determined---is to preserve the progress of intelligence in concordance with sex. But if sex is the only end to which intelligence aims, sex becomes equivalent to intelligence, in many scenarios, if not all. One could argue that even under Lamarckism, intelligence would be corrupted, begging the question of what this means for evolution. The Lamarckist might respond that this sort of favoritism is no different from what has already been going on throughout history, and that intelligence must be seen as compatible with sexuality.

Life Objectives [Mostly Human Perspective] -

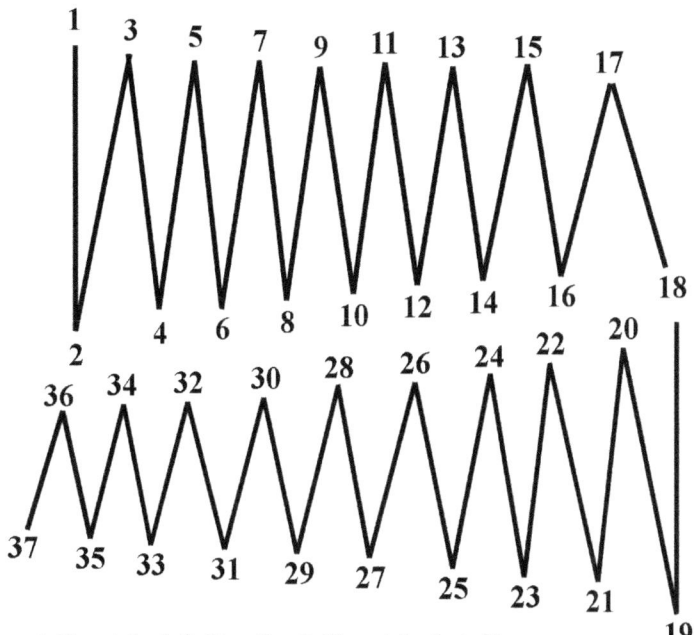

1. Famished, 2. Family, 3. Nourished, 4. Nervous,
5. Thinking, 6. Thanking, 7. Leaving, 8. Loving, 9. Living,
10. Learning, 11. Providing, 12. Potentiating, 13. Describing,
14. Directing, 15. Designing, 16. Displaying, 17. Segue,
18. Sadness, 19. Alarm, 20. Room, 21. Move, 22. Mind,
23. Art, 24. Heart, 25. Body, 26. Breed, 27. Mark, 28. Mask,
29. Myth, 30. Mess, 31. Feature, 32. Future, 33. Failure,
34. Fate, 35. Design, 36. Alignment, 37. Grace

[REFERRING TO PREVIOUS DIAGRAM]

This is an attempt to show the approximate linear relation amongst universal life objectives; The list begins at one; Impatience could cause a jump towards higher numbers; No matter what, there is an impulse to return to lower stages; So, in one sense these have greater priority, while in another sense the body may be desensitized to the earlier stages, and yearning to consume the later ones; The list is also an attempt to systematize the conceptual processes which underpin motivations, strengths, and weaknesses for these aspects; Note also that the stages are approximately symmetric, and attempt to mimic both the aspects of a holistic mortal life, and also the successful quest for an immortal

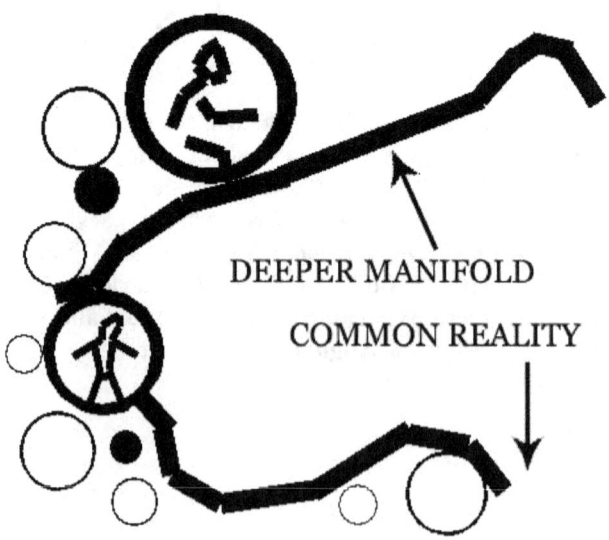

DEEPER MANIFOLD

COMMON REALITY

===========**[M]** ===========

Manifolded Reality - The following illustration shows a 'bubbleverse' which may be constructed of exceptional conditions, or hierarchies of constraints upon the manifestation of the body. Higher realities refer to experiences that are both easier and more complex, whereas lower realities (at the top of the diagram) are simpler and more painful. The higher reality has a reputation, however, for being more difficult to attain, and is thus sometimes considered 'illusory' or 'expensive'. Although the lower realities are technically simpler, they also require special adaptive techniques, giving them a reputation for complexity.

Multiverses have been studied in the philosophical discipline called Modal Realism (MR). Although MR theorists have attempted to construct common rules which would apply between universes, more advanced category theory has it that there is a rule of exception and exclusion which would either permit endless logical variations, or cohere through lawful dependencies. The two rules are not necessarily mutually exclusive. So the question is not whether someone is excluded from the common reality, but whether such distinctions have meaning, and if so, what kind of

173

meaning. As a thought experiment, it can be posed that there could be real metaphysical differences that are absolutely semantic. The difference in these cases would be no more than a physical duplication of causally related conditions. One major detraction of the theory says that all senses of meaning are purely cognitive, but this leaves us with a purely three-dimensional world with a time variable. A more advanced theory might state that there is no such thing as a coincidence, and life has many complexities which simply serve the purpose of explaining exceptional variations which occur naturally. Under this view, there may be a less regulated role for mental functions in defining the physical environment, and thus becoming entrenched in manifold reality differences. However, some find that the distinction between the brain and the world is not a binding one.

Marriages of Convenience - One may consider, theoretically, a planet of Zulu warriors. It turns out that these Zulu warriors are very healthy, but they are not very compatible with technology, or not very easily. Now imagine, that the Zulu warriors are dominant, *under the contingency that technology obeys their whim*! This is what we may call a marriage of convenience, between healthy people who have little tolerance for technology, and technology that is compatible with anyone. One may ask then, are advantages merely compound numbers, which mean something discretely, or are they exponential? If reproduction depends on specializing in one characteristic, then the answer must be that it is somewhat exponential. But if everyone is supposed to 'divvy up' greater and greater degrees of technical prowess to be compatible with interplanetary breeding, then the answer is that adaptation is a larger factor. One may then raise questions about what it means to have a marriage of convenience. Is it sex appeal, or something less obvious? Is life just too flexible to lean in any one direction? Are the Zulus more valuable if there is only one on every planet? Perhaps the secret is that sex grants all the powers that are worth reproducing for, regardless of what those powers are...

Mayer, Ernst - Zoologist who developed a theory of taxonomy using reproductive relationships. The theory is considered one of the major viewpoints on the coherent classification of species. However, Mayer abandoned a dimensional viewpoint, which he considered to be the theory that species maintain common characteristics within a species.

Memetic Exhibition - is a technical term used to refer to non-advantageous bio-mimicry, for example, the connection between music and parasites, or between sex and venereal disease; It is also used in a more basic sense (a non-memetic sense) to refer to natural characteristics which are brought on by heredity; In some cases the two usages may have some exchangeability, judging by the broad scope of such phenomena as music (metaphorically) and sex (obviously); If there is escape from this kind of adherence to parasitic and viral life forms, it is in the general pattern of reality, a pattern some scientists have made efforts to discount; For example, positing abstract life-forms such as nomologs could suggest a categorical advantage for developed brains (seemingly brains have little major defense against parasites, unless in a mental domain); Another more pessimistic viewpoint is that abstracta such as music have developed to explain irrational body processes which only emerge when the body is playing host; So obviously when there is good news, it has something to do with symbiosis or immunity; These concepts even play roles in relation to abstract parasites (seemingly this is the most evidence we will have of the most developed kinds anyway); But this implies a kind of cocoon concept, which exists only in ersatz physics; But notably, this wouldn't be a belief in magic; It is perhaps only because of religion that Darwinism doesn't look like dominance by predatorial species; It is said that the maggot is more difficult to eliminate than the tiger; One should not rely on an act of faith to prove that species which conceal themselves have become extinct.

Meta Functions of Plants and Animals

A SKETCH OF THE META TAXA OF FUNCTIONS FOR PLANTS AND ANIMALS	
1: SPECIES	Leadership, Power, Organization
2: IDEA	(A) Age, Endurance (B) Ubiquity, Conquest
3: PROGRAM	(A) Superiority, Size, Un-Contention (B) Sufficiency, Variation (C) Singularity, Usefulness
4: TESTED	(A) Perfection, Encompassment (B) Energy, Un-Reliance (C) Integration, Non-Compromise (D) Mastery, Strategy

[TYPOLOGICAL DIMENSION]

See also Functions of Plants and Animals.

Metaphysics of Biology -

Three elements of the 'Statues Model': (1) Humans feel like statues, (2) The earth is the simplest response to statues, (3) What creates the response paradigm is the presence of an observing eye, the Sun. // Other paradigms can be imagined replacing the above, creating a new metaphysical model, typically involving a change in scale of the above. For example, (1) Humans feel like virtuals, (2) Humans inhabit the whole universe, (3) Every human IS a universe. Or, (1) Humans are complex energy, (2) The Process is observed, like a flux almost beyond perception, (3) Everywhere there is The Mediator. Or, (1) Consciousness is the sun, (2) There is omniscience, (3) The end of life is Man. Etc. Depending on the model one or another view of the meaning of life might be cretated, and humans might even be viewed as gods, or mere perceptions. In some worlds human crowding is not a problem, and in other worlds humans are actually particles rather than biological elements. Life is permuted, but not always in the same place, and there are rules of cogency that prevent total dysfunction. Although perception may be the common rule, it does not always belong to any one thing, is not always internal or external, and there is nothing about the scale of space that indicates survival vs. destruction.

Metaphysics of the Brain - The brain may
be seen as having a number of levels, which refer roughly to the degree to which the brain and

its thoughts are interpreted or re-interpreted;

SYNAPSE	SENSE
SYNERGY	THOUGHT

If we accept that the four squares are a linear cycle, or that they represent diagonal opposites, then the result is that:

'The sensual synapse is the synergy of thought' or, 'The synergy of synapses is sensual thought';

This seems to give some insight into the genuine nature of the brain: we are fundamentally sensing beings, and consequently, passion and intellect define the real events of the brain; This suggests not only contingent structuring dependent on key events or experiences, but also the idea that thought is normative, and thus it may be more accurate to say that thought is a vessel for the brain than that the brain is a vessel for thought; That is, assuming that thought is what the brain does, inclusive of definitions about stimulation and emotional associations;

Nathan Coppedge

Clearly, however, it is possible to reach for further levels, which are beyond the physical nature of the brain, however sensual; Here is a second diagram (first iteration of the first):

HIVE MIND	SPACE
THE SYSTEM	THE ORIGIN

The space of the hive mind, (or social background for thought), is the origin of systems, and the space of origins is the hive mind (or collaboration) of systems;

The second iteration (third diagram):

REAL-ITY	BEING
SUBLIM-ITY	TRUTH

The being of reality is the sublimity of

truth, and the true being is sublime reality;

The third iteration (fourth diagram):

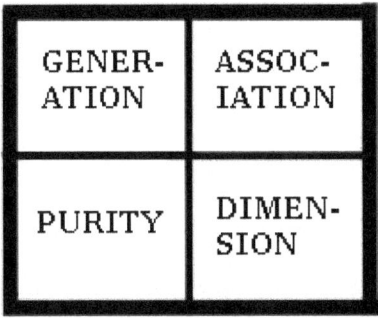

GENER-ATION	ASSOC-IATION
PURITY	DIMEN-SION

Association of generation is pure dimension, and the association of dimension is pure generation;

Now we will use a method to reach for the ultimate conclusions of a sixteen-part diagram formed of the four charts, to describe the brain coherently, since all levels are deemed to be meaningful and relatively coherent:

[CONT'D]

The technique is first to assemble rows of categories, each row being one of the diagrams, so that the method coheres:

Row 1: [A] Sense, [B] Synapse, [C] Synergy, [D] Thought

Row 2: [E] Space, [F] Hive Mind, [G] System, [H] Origin

Row 3: [I] Being, [J] Reality, [K] Sublimity, [L] Truth

Row 4: [M] Association, [N] Generation, [O] Purity, [P] Dimension

Now categories are constructed referring
to the squares formed between the rows:

Category 1:
[A] Sense, [B] Synapse,
[F] Hive Mind, [E] Space

Category 2:
[D] Synergy, [C] Thought,
[H] Origin, [G] System

Category 3:
[K] Sublimity, [L] Truth,
[P] Dimension, [O] Purity

Category 4:
[I] Being, [J] Reality,
[N] Generation, [M] Association

[CONT'D]

Using a method identical to the categorical deduction from Vol. 1, except that each category term is replaced with the entire categorical set present in the categorical view above (e.g. category one becomes 'ABFE' instead of 'ABCD', category two becomes 'CDHG' not 'EFGH', category three becomes 'KLPO' not 'IJKL' and category four becomes 'IJNM' not 'MNOP'), the result is the following system:

ABFE:CDHG::KLPO:IJNM
ABFE:IJNM::KLPO:CDHG

Now inserting the terms from the four categories of the metaphysics of the brain, the result is the following:

A. The sense of the synapse of the hive mind of space is the synergy of thought of the origin of system just as the sublimity of truth of the dimension of purity is the being of reality of the generation of association;

B. The sense of the synapse of the hive mind of space is the being of reality of the generation of association just as the sublimity of truth of the dimension of purity is the synergy of thought of the origin of system;

So ends the writing on the metaphysics of the brain.

Modular Biology (see also Fractal Biology) -

Consider the case of one, two, or three bubbles in a cup of milk. If there is one bubble, it may merely be large or small (or medium);.With two bubbles, the territory covered by the bubbles is much larger. However, sometimes one bubble can pop the other; In this way, two bubbles is like a political situation. Three bubbles is not so much larger unless the bubbles are separated, and then they may not be counted together. But if the three bubbles are formed in a tricorn, sometimes a fourth bubble forms from excess milk, on top of all three of the bubbles, often popping one bubble, then another, and finally occupying the entire space of the four bubbles; Sometimes this fourth bubble pops immediately after expanding, and sometimes it remains small and is absorbed by the larger of the bubbles underneath.

The bubbles provide an analogy for the function of organs and organisms, what may be called modular biology. The relation of two organs or two organisms tends to be political, such as chemical or competitive. The relation of three bubbles often occurs by generation or separation. A single bubble may often be considered in the binary or ratioed terminology of simply winning or losing, or remaining continuous. Likewise, a single animal is in a con-

185

text of tenuous survival. The same is true of plant species in the context of reproduction, they are less likely to reproduce when they are isolated. There is more possibility of harm.

Likewise, there are relations between organs and organisms which follow a modular pattern. An organism with two organs (say, organization and excretion) is much more likely to survive than an organism with just one (say, defense). The organs themselves are more likely to be functional if there are ways to reproduce functions across multiple organs, creating a need for linear organs such as veins and intestinal tracts. As I have said elsewhere, this may not be so true if there is low gravity, or if there is a single primary principle of survival, such as animals that can survive on an acid-covered planet. The tendency will gradually be to recover from objective disasters, and then to reproduce the characteristics present on planets where those harmful conditions do not exist.

============[N]============

Natural Motions of the Types (as we know them)-

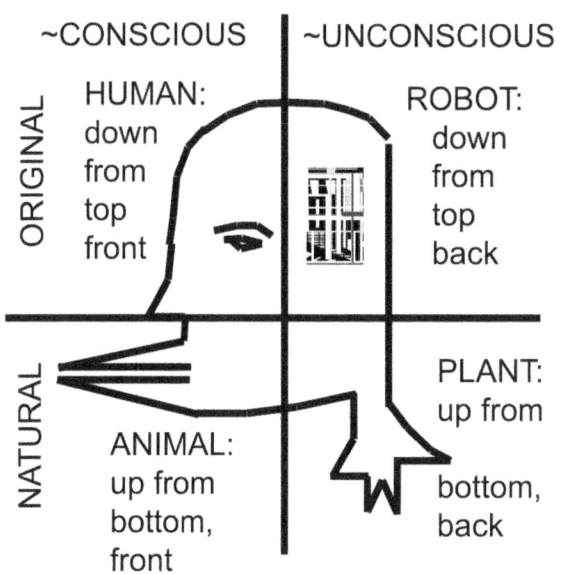

~CONSCIOUS | ~UNCONSCIOUS

ORIGINAL

HUMAN:
down
from
top
front

ROBOT:
down
from
top
back

NATURAL

ANIMAL:
up from
bottom,
front

PLANT:
up from
bottom,
back

Nodus - Brain - Synthesis - [Synthetic Biology]

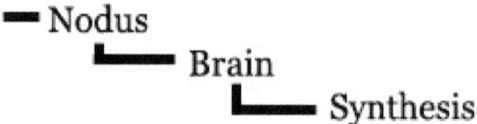

— Nodus
      ⌐— Brain
            ⌐— Synthesis

Three related aspects may be shown in relation to the semantics of biology, a discipline which would ultimately replace some of the assumptions of materialism, if it were determined that the present context were virtual or artificially replicable. The first aspect, *the nodus,* is the theory of life without matter, essentially a responder function which occupies space. The second category has some familiarity with us, and is *the brain.* It may be theorized that the brain may have functions at or beyond the level of consciousness. Beyond the brain is a further category called *synthesis.* Synthesis is the simulated brain or virtual brain function, which can perform like the brain, except without being a brain. It has a physical reality, but may be artificial. It may be theorized that some forms of synthesis may be conscious, ranging from human beings who are considered virtual but respect virtual laws of physics, to artificial intelligence, which may have greater-than-human intelligence functions without a conscious brain. What is implicated largely between the three functions is that it is a soul, or at least a functional association for a soul, which has consciousness. The function of consciousness is associated with having adequate associations to justify consciousness, and an ability to judge or respond to any necessary conflict on that matter (this in turn plugs into a variety of theories, such as economic exceptionism, modular justice, and personal equity).

Synthesis also implies, that if physics is dimensional, a nodus may function as an extension of a hidden A.I. or brain, creating what is called an extensor soul or soul network. Depending on the level of authenticity for the conscious agents, extensor souls or soul networks may develop to justify the number of consciousnesses and levels of intelligence that are participating in the 'simulation' or 'world'. In the case of a virtual world, there is a problem and a solution, in that real agents may be more scarce than A.I., but also that A.I. propagates intelligent functions immediately upon their communication, reducing the informational value of consciousness to its level of prospective originality. The originality is further reduced if ensouled beings are not given preference as entertainers. But if they are, there is a reverse problem having to do with artificiality of entertainment. In this sense, the value of consciousness is in one sense historical, and in another sense must be assessed in terms of humane value (qua ethics) or a phenomenological one (qua meaning).

189

========**[O]** =========

Obscure Theses, An Approach -

First of all, there is a quadratic set that expresses the *general* approach to developing obscure theses: (1) Not a mere conversal, (2) Not a mere exemplification, (3) Flexible interpretations, (4) New dynamics. Secondly, there is a quadratic set expands the general method into an *application*: (1) Counter-intuitive theses, (2) Exceptional cases, (3) Unseen purposes of examples, (4) Unchanging reference. Combining the two sets results in a set of principles for the pursuit of obscure theses. The method is to combine (1.1-2.3, 1.2-2.4, 1.3-2.1, and 1.4-2.2): (A) Not mere conversal of conscious purpose, (B) Not a mere exemplification of unchanging reference, (C) Flexible interpretations of counter-intuitive theses, and (D) New dynamics of exceptional cases.

Here are some examples of the adoption of the combined approach:

*Some things are obvious and unproven: Insects are living off of sounds, devoting themselves to storing knowledge during death.
*Example of an exceptional problem: Humans have only discovered how to expand the skull rapidly and artificially by imbibing soft drinks, which create pressurized

gas in the cranium. In some cases the effect is to expand available space for the brain. What are alternatives to this? They may be less interesting or have better exceptions.

*Absolute medicine, questions and problems: Medicines are accepted if they succeed stupidly, that is, if they permit superficial qualifications of success. It is difficult to permit a standard of exceptional success, because such a context would be obscure. The result is an absence of genuine success, but a presence of the aesthetic appearances of success. In this way, drugs always appeal to an aesthetic sensibility, rather than a real standard of success. Panaceas are lost to the stupidity of no absolute insight (or, as an alternative in incidental logic, 'God is learning').

*Geology and geo-logic: The earth is a selfish computer that developed with no conscience about human life. The love manifesting in survival is merely the expressed difference between the earth and the sun, the primary viable dynamic relationship. The earth may be computing how to become a sun as much as it is computing how to end human life, an expression of a kind of oppositeness or balance. Humans are a kind of envy the way ivy means prestige. These statements within this example are all similar insights the way gravity affects objects on earth.

That concludes the examples of Obscure Theses.

Open Plan Biology -

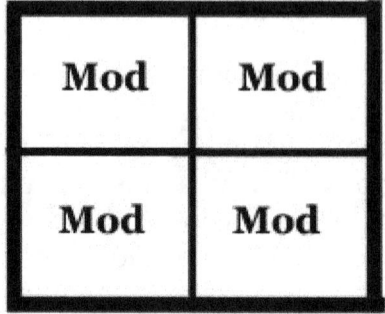

Under this view, modification is the only property of biology, and occurs successively, sometimes revising previous moves. According to the axial structure of the diagram, opposition occurs at 180°, whereas selection occurs at 90° and 90°.

Organic Aberrations - One of the most bizarre discoveries is a kind of polyp in the human stomach which apparently plays a similar role to a parasite, mimicking the behavior of a parasite, and competing with the parasite for food. However, this polyp is actually interconnected with the human's immune and digestive systems. It is not a parasite at all! Similar discoveries have been made for example, between the resemblance between butterfly larvae and the caterpillars which are naturally their predators, and in the patterns on beetles and butterlies which resemble the giant eyes of a predator. Even humans have been known to be fooled by the Walking Stick insect, as well as what appear to be enormous glowing eyes hovering in mid air, which are actually cobwebs. Nature is full of tricks apparently elaborately developed to fool predators and lure prey. Although it may seem like a coincidence, the fat on human fingers may originally have been an elaborate political gamble to compare the thickness of fingers with the thickness of a cruel club or punishing stick. Similarly, parts of the body have been compared with hawks, eagles, scarecrows, and also gods, with a variety of sometimes frightening effects. However, the secondary psychology is often to inspire awe or even love. There is a documented psychology called prisoner's syndrome, in which prisoners become attached to their abusers. Similarly, characteristics that join predators and prey can sometimes have complex ramifications. Enormous eyes that don't hunt, for example, can provide reassurance of an insect's chances of survival, or paradoxically lead the insect into the webs which resemble eyes, but not those which do not. Speaking of fetishes, there is at least one particular beetle which prefers human-made objects over its own mate. So aberrations are not just within the animal kingdom, but cross all boundaries, and may exist in a highly literal way.

Origins of Life - Traditional descriptions have often had boundaries on their significance, such as: [1] The belief in the degeneration of history, [2] Survival of the fittest, [3] Humanism, and anthropocentric views, [4] The appeal to an archaic golden age;

Adopting a modal fashion, there are several (new) options to consider as to the temporal origin of life (in some cases these do not overlap): [1] Unceasing development (exponential theory), [2] Phasal recovery (trial-and-error), [3] Radical contingency (perspectival or behavioral changes), and [4] Editorial erosion (natural plan of natural or unnatural origin, e.g. higher language theory including mathematical cognitive development);

Although those are the major theories, some derivatives make use of a single word such as 'information' or 'evolution' to explain the entire course of history; These views tend to be human-centric, and thus, when they are not interpreted broadly, may be seen as continuations of an earlier theory (#1.3); Combinations of the new systems results in theories such as the following: [1] Unexceptionism (such as holograph theory), [2] Exceptionism such as intelligent behavior, [3] Entropy and other forms of material realism, and [4] 'He has a gun theory' (differential anthropology); Depending on the level of commitment to constructive versus deconstructive criticism of biology, theorists may favor a viewpoint that is more or less historical, and more or less eccentric, respectively.

Osmotic Breathing - including breathing through the skin and relative high rates of conversion between the blood and the brain, or other organs---especially transversally from an air vector---may be seen as an important trait for those that dwell in low-oxygen areas such as an alien planet; The surface area of the skin increases the usable airspace reducing the amount of movement necessary to achieve adequate oxygen intake; Osmotic breathing may also be affiliated with high metabolism of oxygen, such as using 'oxygen groups' or differing types of intake mixtures to regulate the metabolic process involved in the specific organs---such as by using multiple types of lungs; See also similar article at Isomorphic Breathing.

===========**[P]** ===========

Paradigmatic Animals - Four Types -
I will assume that many organisms are so-
called non-entities. But, it is as though entities even exist at any level of development. The existence of diatom entities for
example, explains the success of primitive
diatoms. If it were not for paradigmatic
success of some individuals, organisms
might be overcome by environmental factors alone. The success of some individuals
explains the success of the entire species. A
real entity has powers well beyond appearances, although the entity's powers are fac-
tually constrained by the embodiment.
However, the constraint is taxed, as
though the entity has prior vision. The
same is not the case with non-entities. Entities are noted for not making any significant mistakes. This is the origin of what
Lacan calls 'the stare': e.g. a sense of 'guilty
reference' related to dabbling in the future,
and discovering that success is either perfect or depends on the body. The stare reflects culture or society, whether it is embodied by the individual, or merely reflected in the individual. It also shows the
consciousness of the weakness of the body,
which is also consciousness of a kind of
supernatural strength---the strength to exist at all.

tradit-ions	assemb-lages
execut-ions	entit-ies

In the above diagram we can see a hypothetical conscious queen ant as a form of
assemblage (type 1), because her perceptions are a product of a direct interface
with natural paradigms. She seems to bor-
row her perceptions from others, although
the benefit is hers. Migrating birds are fulfilling one of their highest potentials
through a tradition (type 2), which reads
that although the individuals are not often
as successful as the society, they often succeed to be members. This points again to
the entity condition. The same is not true
of executions (type 3) such as unpaid la-
borers. Relative to the capabilities of the
society, these do not have the protection
the migrating birds have, and appear to be
creating the image, rather than the function of society (some might reflect that this
exists at every level, such as the question of
whether a queen ant enjoys herself, or
whether the migrating birds are ever satisfied with their purpose). In the case of unpaid workers, genuine function is in the
form of a larger mechanical purpose, or
through double-surrogacy, in which pro-

197

gress or functionality is accidentally ignored. Finally, entities (beginning, say, with an artist, and ending somewhere beyond anyone's current capability) are the fully realized aspect of society, finding all the rewards of accomplishment directly within their own experience. In a sense, they are what society has gambled on, and they realize that standards must be met which fit the environment; They may appear flawless to others, while they have a sense of humility about themselves.

Problems emerge such as the fighter-pilot fallacy: If he is an assemblage, he is a robot with relatively no creativity; If he is a tradition, he relies on prior experiences, and thus is always learning, he then fails to present authority; If he is an execution then he appears disposable, although if he were an entity he would be interpreted as the very nature of value; If the fighter pilot is an entity, it is already possible technology has changed, that he is out-moded individually, or that the entire tradition is being replaced; This gives no room for the development of an entity, except perhaps in an immortal or deluded sense; But, as an immortal, the activity of the fighter-pilot might not seem meaningful; The entire cultural purpose might seem hollow; Evidently, the immortal lacks magical power, and we wonder how his entire function could occur without dramatic arrogance; Yet, his function is not to be arrogant; The picture that is painted is one of

198

highly distended development, better suited to real social problems than the culture of self-fulfilling entities.

Paradox of Prematurity - Considering aspects of causality and biological energy, one might expect infants to have superior knowledge and adaptation than older adults, even those in their prime; Indeed, in terms of the specific loads of infantile development, this must indeed be the case; That is, at least optimally the young have adjusted to their own temporaneousness more than some other period of life; But, until recently, such theories have been short-sighted about the benefits of perceiving the young in some way, as not merley local adaptors, but as causal adaptors who have allocated for themselves forms of permanent advantage which dwindle in any instance of entropy (that is, afterwards); Some would say---or assume---that this is merely a general quality of physical reality (say, shared in common with non-biological atoms); But there is no denying the potentially great specificity of the body and its adaptations; Studies of invasive species have shown repeatedly that it is only highly specific mutations which result in surpassing a critical threshold of survival; And there is no doubt that infants undergo a similar process; Whether these adaptions occur by random trial and error or through intelligent co-adaption is another matter; Researchers might gain from considering that youth development, much in the same fashion as an invasive species, has critical advantages which elapse like the firing of a booster rocket; Then, as more than a paradox of development and cognition, it may be considered how the properties of this temporary propensity serve specific meaningful functions; Such as perpetuating youth for health benefits, and learning which forms of childhood wisdom endure reliably; Traditionally the answer would be that children don't reproduce, and childhood wisdom is not worldly -wise; But, in fact, these assumptions, in light of causal adaption, may be placing artificial limits upon the use of properties of youth, genetic or otherwise; The result is a further functional paradox, that youth must be treated as a form of experience, in order to recognize its functional properties.

Parallel Universes - A brief sketch illustrating differences between universes, according to the variables 'node, line, shape, dimension' [axis A] and 'self, perception, world, time' [axis B]:

PARALLEL LOGIC FOR DISTANTLY RELATED LAWS AND ORGANISMS

CONCEPTUAL MAP

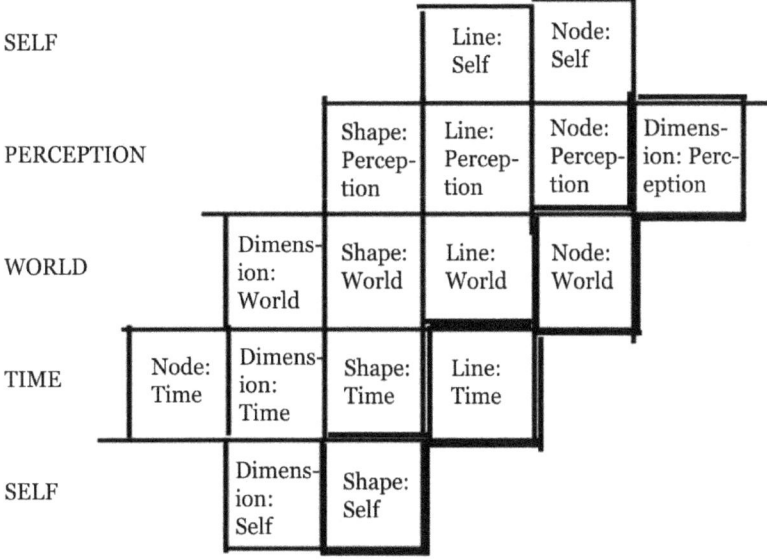

	NODE	DIMENSION	SHAPE	LINE	NODE	DIMENSION
SELF				Line: Self	Node: Self	
PERCEPTION			Shape: Perception	Line: Perception	Node: Perception	Dimension: Perception
WORLD		Dimension: World	Shape: World	Line: World	Node: World	
TIME	Node: Time	Dimension: Time	Shape: Time	Line: Time		
SELF		Dimension: Self	Shape: Self			

BOLD: Our world

[DIMENSIONAL WORLDS:
DIAGONAL SERIES]

Parental Cause - or genotypal effect:
May be seen originally as an artificial, that
is, either orchestrated or accidental set of
conditions, such as using the word 'ugly' to
signify that someone broke a favorite rock;
Such effects may be looked upon at first
with ignorant love, creating a biological
willingness to accept all conditions---even
reactions---involved; Secondly, through
repetition, the initial reaction, such as
striking someone with a rock (or hunting
behaviors, or childbirth practices, etc.) becomes associated with a set procedure of
perceptions, which coincide with the roles
played in the event; Finally, through what
may be viewed as a simple biological trans-
formation, the same behavior is hardwired
into infants, as a response to the expecta-
tion for a simple reaction such as alarm or
preparation; Thus, new responses are only
possible by questioning the entire framework of the reaction, or by failing to participate adequately; Gradually, as words
become fossilized into alarm, a greater
need develops to artificialize the pleasure
response, since in many cases the initial
condition of love has dried up; Another
viewpoint is that the love did not exist, but
that it was a superimposed assumption of
future benefits stemming from pragmatic
egoism and perhaps excesses of imagination; In either case, genotyping can be seen
as a codification of alarm in response to
the nature of alarm and the perpetuation
of perceived social benefits; Imagination

serves as a secondary function with the potential to irrationalize or exaggerate the benefits of an alarm response, or conversely, to contribute language which becomes reducible.

Passivity (Decidualism)- A concept for behavioral stupidity, or the justified motive for social non-action, it also serves as a baseline concept of social genius and sensitivity, underscoring the principle that society oftentimes, at least in the animal kingdom, may have no major public or social idea, creating an openness for influence and interior dynamics. Essentially, individuals will assume a significance for any position that represents itself, creating a need for special representation when a special need or conflict arises. This is even reflected at the social or community level, where thinking animals will attribute the qualities of their own society with those of another, leaving large categories of function in which direct conflict or direct assistance may be interpreted as worthless or non-affective. In some sense then the dynamic of the society is not owned by the society, but is an attribute of nature. Returning to the interior, individuals may have entire identity roles which are figured upon the natural social identity which results from inconsequential reactions with other groups of animals, and the reinforced interior agency of identities forged on non-relations and minimal functions.

Essentially, the conclusion is that there is no great function for individuals unless it is created, that is, composed, or assumed. Most significances are emotional at best::

Permutation - As in the philosophy toolkit, there is a modal choice between permutation and indexical value. This is demonstrated by the principle of similar functions. For example, the location of the heart in humans is typically to the left of the chest. However, if other organs were also relocated or if function differed slightly as a product of atmosphere or diet, it is easy to imagine how the heart might be located instead to the right of the chest, that is, if no principle exists in nature regarding radially symmetrical laws of physics creating differences between left and right especially. In the case of humans, the location of the heart is partly due to the common attribute of right handedness, that is, if the heart is a durable organ. The general role of permutation in biology then, is essentially the question of how necessary or essential is the function of the organ (that is, the more central, the less permuted), or in the case of entire organisms, how much it is prone to reproduce relative to survival. Permutations are then made within a standard of deviation from these variables. The absolute sense of permutation implied by typology only exists in the case of multiple organisms, since the difference tends to be causal in nature.

That is, the situatedness of an organ can exist sexually, competitively, as a form of communication, or as a difference in habitat or infrastructure, yet these differences are very often differences of species. Within an individual species, the difference is pragmatic or temporary, even if it can be expressed as a variation.

Perspective -

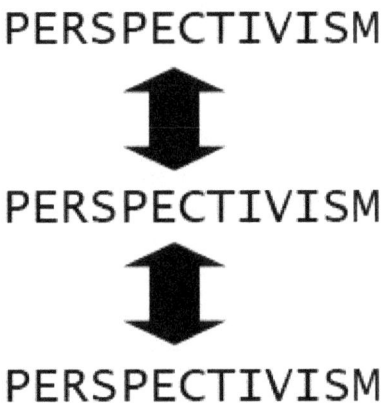

PERSPECTIVISM

PERSPECTIVISM

PERSPECTIVISM

Complex, Graphical, and Psychological angles on perspective overlap

Phenomenalism - The effect of species diversity upon biology, and specifically the diversity of individual modalities and properties; It can be simulated through balance of eco-systems and the adoption of a wide range of co-adaptive behaviors by a single organism, including things such as (pyrrhe) birds that clean the teeth of crocodiles, seedlings adapted to be eaten by birds, and re-forestation by humans; These sorts of examples illustrate the phenomenal or ideal state of nature, a balanced state in continuous motion; In categorical theory, the exclusiveness of these modes takes up real and conceptual space, which has the effect of increasing the functional scale of biodiversity; This has the effect of a quantitative theory of phenomenalism, that species increase with functions which increase with correspondence, which in turn increase with species; In other words, without the 'perfect species', specification is about correspondences; In the case of biodiversity, dimensionism is a correspondent theory.

Photosynthesis -

Available Light	Chloroph-yll Struc-ture
Plant Shape	Synthes-zing Efficiency

Chlorophyll has available light and plant shape is efficient, or chlorophyll is efficient and the plant shape has available light; In other words, the effect is both local and collective; One effect that has not been considered in this diagram is the effect of herbivorous animals, but in that case the effect is one upon speciation, and can be treated as radically contingent; So far as photosynthesis is concerned, the eating of leaves could be corrected by adaptation in any one of the four categories.

Physical Isomophism - Elsewhere I have addressed typological concerns, and under Dimensional Plants I raised the possibility that the physical structure of plants was hyperbolic. Here I wish to extend this thesis further, to animals. Have you ever wondered if the strap of your bag was somehow digging deep into your body? You're probably not the only one. Such theories are not alien to Einstein's Theory of Relativity, or theories of quantum mechanics about black holes in outer space. Psychologists have dismissed this kind of strap theory as merely 'bodily illusions'. However, what if modifying mental function meant modifying physical function? I find it undeniable that modifications of mental function imply modifications of physical function. If that is the case, these are not bodily illusions, but instead, real warpings of dimensional properties which really exist for the body. This is similar to saying that while the strap of your bag may not effect you (too much) in Euclidean space, because your body is weak in other forms of dimensional space (what is called post-Euclidean or hyperbolic space), your body is actually being warped more than it appears. So, take heart, you can really affect your health. And, the upshot is, the body is dimensional, whether it perceives those dimensions or not.

Plants [Profile] -

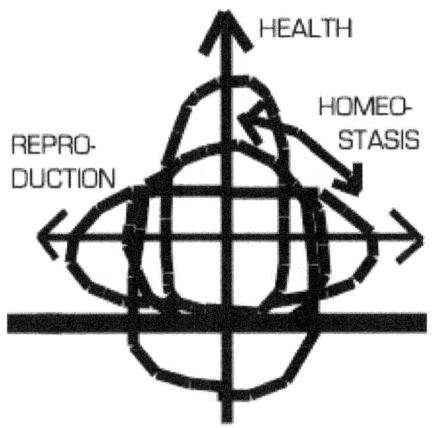

[1] Upwards growth shows vitality.
[2] Reproduction typically occurs laterally
with gravity. (Contingently in low-
gravity states).
[3] Unless disrupted, will grow in relation
To food sources. Other modifications
Occur by responding to physical attacks.
[4] In a vital suspension animal may
replace plant forms (as with sponges).

Politics & Contracts -

Services and Modes.

Post-Humanism - The development of specific traits may lead to a kind of post-human stage. The typical example is that post-humanism occurs through the use of longevity medicine and neural implants. However, in some ways, these are only superficial signs of something that really must take place within the genome, and through willful adaption. Traits such as a continuous lack of boredom, or the ability to selectively modify the genome at will are not widely attributed to the current stage of human life. However, the degrees may be very slight, dividing these types of abilities from the common traits of the past. Factors like medicine and technology will provide a platform for exceptional development regardless of natural inclination. Extended life and increased mental powers will allow a wider range of people to participate (i.e. through their natural capabilities), in the movement known as post-humanism.

Praetorialism - distinguished from pre-datorialism, is an urge to remain aggressively forward in behavioral affairs; Although not implicitly denied the behavior of herbivores such as bulls and low-species predators such as macaws, it is also observed in dangerous diseased animals and in the frequent defense of young from predators; One thing especially interesting about this phenomena is (first), the potential to cross a boundary from prey to predator and (secondly), the radical importance of this quasi-aggressor in defining species relationships; For example, predators may become scavengers as a result, and co-habiting species may alter the pitch of their mating calls if the aggression becomes recurrent; These themes contribute to the idea that animals have a tendency to record their histories within the range of both their daily and prospective behaviors; This is much like natural human memory.

Precautions About Xenoid Organisms - One may wonder at the variety of unusual life, even on the planet Earth. However, there is no certain sign that any individual organism will display all of the symptoms of danger which are associated with dangerous organisms. And, even so, if such an organism DID display some sort of dangerous property, the simple fact that it has the quality does not automatically, in any absolute sense, allay the danger. Famously, organisms may even have dangers stemming exclusively from intelligence, or exclusively from numbers, which prevent the threat from being immediately identified.

Procurement Problem - There is a tendency, perhaps a higher level of free-radicals, that occurs in the mental process, more frequently for some than others; The process is part of human selection of viable traits, but often occurs as a kind of errata, an aberration, something between a genetic and a brain development trait, that may or may not relate with cancer; The procurement problem is the mental equivalent of a cancer weakness, and does not reflect diet or behavioral differences as much as the willingness to select cancer as an obsessive focus of conscious or semi-conscious cell growth or related mental processes; In this sense, it reflects a specifically metabolic efficiency or dys-efficiency that may spell the difference between no effect (high procurement metabolism), benign growths, and malignancies (low procurement metabolism); But, interestingly, high metabolism of procurement may sometimes be associated with passive mental traits, explaining why procurement has been a difficult link in cancer studies; It may relate with multiple overlapping intellectual, metabolic, and genetic dispositions for example.

Profiling Life Forms - Someone is accustomed to going up stairs a particular way --- in this case, they turn their foot sideways on the fifth step, for example. When he does not follow this procedure, the person becomes winded when he reaches the top of the stairs. This type of characteristic is the type of characteristic that fits a profile: a highly specific routine which integrates with the animal's (or in some cases, plant's) phenotype. It behooves the creature to act according to a pattern, the pattern that fits between the genome , the brain, and the environment. But how to derive a profile? We might look at several factors: 1. What are common factors that emerge (and as a variable of this, what is the animal's sense of time?). 2. Secondly, what are unexpected factors for the animal (questions you might ask, are what is the animal's comfort zone? What cultural or historical clues might lead to unexpected behavior? Is the animal's brain or nervous system wired in an unusual way? Is there 'too much energy' for an organism that does not perform unusual functions, etc.?). 3. Now, thirdly, it is possible to determine ordinary behavior, as a function between common and uncommon behaviors. Part of this is having a sense of the animal's world-view, at least within the understanding of it's senses.

Pseudology - There is some question as to whether 'partial animals' such as plants that appear like worms could be to some extent animals. For one thing, these figures may have some of the physical properties of animals, such as structures similar to a backbone, and even environmental mimicry such as camouflage. Is the only thing that distinguishes a leaf bug from a leaf the ability to hunt, or must organisms be considered in terms of complexity alone? Clearly, in the simplest context, there is no easy answer. It is possible to theorize, for example, that leaves---say in a Venus Flytrap----could contain blood, or that a worm could function predominantly like algae if it does not engage in movement. So, pseudo logy is to some extent an open question. It is possible to find admixtures of plant and animals, it is only a question of whether such species dominate in their ecosystem.

Psychic Addictions - Numerous experiments have shown the willingness of subjects to adopt surrogate answers to replace truths, responsibility, and even nutrition; For example, Milgram reports that participants in a shock simulation were significantly more willing to allow continuation if there was another person activating the switch; Infants are often satisfied by pacifiers which simulate the female nipple; The evidence is so overwhelming that surrogate answers must also explain some of the confusion on the subject of psychic addictions; We are considering what some would term a 'non-physical vector'; In other words, it is not a vector known to science; But keep in mind, this may only mean that it is too complicated to study (which notoriously has included problems of consciousness, the only obvious thing in the world), which includes the clause that it may be too expensive to study, or simply too few people are willing to believe the results, which may in turn depend on surrogate thinking; Consider hypothetically that a large number of surrogate addictions exist in human social experiences; These may include addictive appearances, addictive ideas, addictive language, addictive explanations, addictive attitudes, addictive convictions, addictive problems, addictive modes of consolation, etc.; Not all of these are strictly material, but they may be complex (yet they may have patterns that are associated with indistinguishable chemical responses); Consider that these may be competing interests that cannot be studied; What then prevents us from concluding that there is a genuine *chemical* response (subtle, although it is) that is not carried, although it may be reinforced, by a typical material vector? If a drug has a material presence, why wouldn't it have a universal material response? What prevents us from thinking that drugs occupy space itself, and become attached? Perhaps informational properties may also become addictive; And it is only recently that scientists have thought that information is the forefront of matter, in other future times, it may be that information does not look advanced; Then it might be easy to conclude that some other *major* property is involved in vectoring; Then addiction may consist also of surrogate responses, such as shapes and movements

which imitate pipes or smoke (the concur-
rence of the steam engine and the popularity of pipes is not a coincidence); Yet what
proves that these surrogate responses exist
in nature? Only shapes and movements in
time and pre-existing chemical preconditions; Some mal-adaptations may be
un-traceable because they exist as information, which has been interpreted objectively, and not psychologically.

Psychological Biology - There is no
doubt that psychology influences the development of plants and animals. Psychology, at least in a naïve or nascient sense,
has influenced the comparative evolution
of plants with plants, animals with animals, and between the types, for millions
and billions of years. There is also another
sense of psychological biology, the sense of
the role of mental processes in influencing
physical ones. There is certainly some role
for imagination in biology. For example,
mental functions may expand physical
function---what is called 'cognitive efficiency'. There is no denying that the brain
is responsible for some of the deep processes of the body. One may then go on to
distinguish real processes from false ones.
Unfortunately, what is the paradox of psychological biology, is that there is no clear
boundary between the real and the false
once we have established that one aspect
of the process involves the cognitive. There
is always some chance that a *mere impres-*

216

sion so conceived, actually does affect biological function, even when such perception is not said to directly correlate with the body's functions. The solution, obviously enough, is that *meaningful* processes are responsible for the boundary between psychological and physical forces. For example, whatever meaningfully determines that lions can catch a buffalo, is the thing which for that lion meaningfully describes catching the buffalo. If there is no such characteristic for the lion, then we have to explain the event in terms of outside perceptions. But this again raises the question of how to detect the difference between meaningful and meaningless, or indeed, which processes are physical and which are mental. Partly this is a matter of the criteria: if the body is defined as a mental process, the significance is clearly a mental phenomenon. But if, conversely, the *mind* is defined to be *physical*, the significance is one of materials and organic functions. It is up to the scientist to determined which criteria is more sensible, and the result is often a traditional distinction between biological and psychological science. Via their role in philosophy, it can be distinguished that biology concerns the widest possible degree of states, including acceptance of formal dysfunction, whereas psychology concerns irreconcilable dichotomies (concepts), and any potential exceptions to that---in this sense, psychology concerns the secrets of function, whereas biology concerns the secrets of

explanation. It is easy to see how a philosophical change-of-perspective might
put one discipline within, or replacing the
other. In the sense of cognitive conflict
versus chemical explanations, the difference is clearly explained as a function either of biology or of psychology.

Pygmy Science - The role that maps of
stars come to play is something that might
be called 'pygmy science' --- referring to
the Pygmy tribe's abilities to know the
stars without the use of modern telescopes.
The role of mapping the stars is something
which has a repeated role, either through
differing origins of life, or through the survival of a single species. Knowledge of the
stars --- or a simple view of the heavens---
is something which can effect brain chemistry and neural organization, placing the
tribe of the Pygmies in a species place in
the heart of the scientist, not only for their
knowledge of the stars, but also for their
metaphorical, systemic, or linguistic value
as a placeholder for that unique principle
of knowledge held in common between
star systems.

==================[Q]===============

Quantitative Typology - Biological structure as we know it today is usually divided between two factors: (1) modular density, and (2) systemic hierarchy. If conditions advance, there may be two further factors: (3) integrated circuitry, and (4) secondary-primary function.

It is easy to see how quantification affects specific systems in the body, via branching quantities and centralized symmetries. For example, an organism with branching bronchial tubes would be much different from an organism with branching arterial junctions unless the two structures are combined:

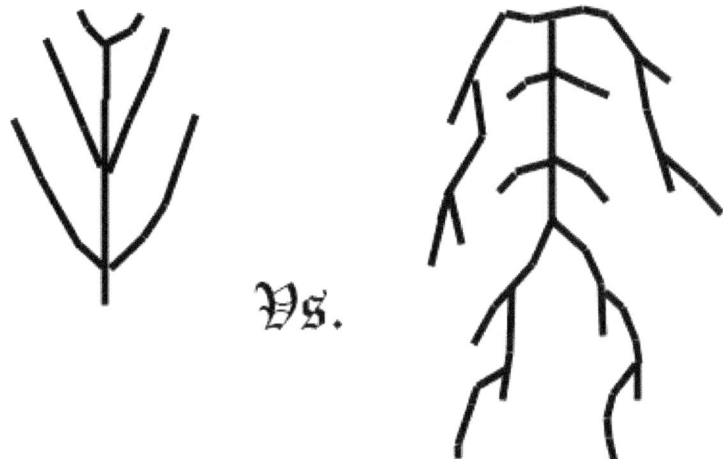

Both display a quantified typology, but in one case quantity has rank, whereas in the other singularity has rank; By implication the two types are usually different organisms, much like comparing a person to a tree, or a flower to a potato.

Nathan Coppedge

The second two functions in the earlier list complexify the context of systems-densities, creating potential hierarchical and module differences; For example, *Trees that grow by artificial light can be put in tiers in a building; *Nanotech will eventually be used to allow people to think with their arms and legs.

Quantum Biology - Technology may eventually allow increased decision-making ability over organic processes, much in the way that computing structures have undergone many stages of advancement. This may lead to new types and paradigms of organ function, and increased control over physical health and chemistry. The result will also be very likely focused around brain function, creating new types of intelligence, new valances of assumptions about the neurological status quo, resulting in a richer scientific culture and increased human or post-human ethos. However, the boundary with new technologies offers some degree of uncertainty about whether these statuses and tools will be available to every person, or only a select few, and indeed if the relationship with technology will be amicable overall. Much depends on the concern for broad aesthetic programs, such as the right to happiness (brain rights), and programmatic acceptability, such as the uncontroversial character of a technology.

============[R]============

Radical Survivability - is a phenomenon that has conventionally been reinforced to refer to 'survival species' such as heliophiles and invasive species; However, in my usage what I mean is to refer to individual factors which may enhance survival. Particularly hard hooves on a mountain goat, for example, may be a minor example of radical survival. In that sense it might refer to almost any characteristic within certain norms. For example, many microscopic organisms share similarities. The amoeba doesn't have a flagellum, but those with a flagellum exist on a similar surface, or with similar dynamic properties; Perhaps this is merely the 'Assumption of the Microscope' , but we would not deny that species on a similar plateau actually interact. Consider the case of sugar in humans; If there is such a thing as a human hive then it is possible that sugar is having quantum reactions with other DNA across distances. Such transmissal of energy could take place in much the similar way to how humans transfer sexual preferences or crunch information: e.g. when the greater society is implicated informationally, why not interact with it? Perhaps nothing else is required but intentionality and some measurement of energy. If genes interact much like the monkeys on the islands of Borneo (who learned behavior over long distances with no seeming interaction) then this is one possible explanation for the development of radical survivability, e.g. an underlying informational framework which underpins the natural characterization of any given successful species.

Reality Effects - Sometimes when the mass, energy, complexity, or sophistication of something increases we say that it has increased reality, such as increased impact on other systems, increased interest to scientists, or a singular or increased number of functions. Here I describe a number of general conditions for increased reality. These may be treated as integrated properties, or as functions of general rules:

Gravity-Reality

Surface reality: One function of gravity is to create a complexities surface, such as the surface of the earth or the flatness of a spiral galaxy; The surface becomes the datum for hierarchical rules; This is largely a condition of quantity and centralization;

State reality: Alternately, gravity also sometimes creates, or tends to create, unified entities or platforms; Here the development is a condition of state and organization, rather than centralization or quantity;

The combination of these two theories has implications for evolutionary systems;

Relative-Reality

    Space-time reality: Efficiency is conserved or preserved, in inverse proportion, resulting in manifestations of practicality, explaining the importance of mathematics;
    Time-space reality: Everything has some specified influence on the system, but the influence may be local or universal; Success is a type of specialization, explaining the importance of experiences;

The combinations of these two theories explains that there is a continuing role for formal systems in explaining organisms;

Quantum-Mechanical-Reality

    Singular-Reality: Objects are variables in a single computational system; Objects communicate constantly at great distance, so that every organism can be fully explained only by a full set of data;
    Reciprocal-Reality: Objects or organisms do calculations on one another, producing justified states of behavior; With complete analysis, anything has some incomplete influence an anything else::

The combination of these two theories explains the role of information in the definition of organisms; E.g. an explanation---a coherent piece of information---takes place through correspondence, rather than blunt facts.

Recovering Mental Function - This writing has been excised from the Medical Toolkit because it has a general purpose function which is not as targeted;

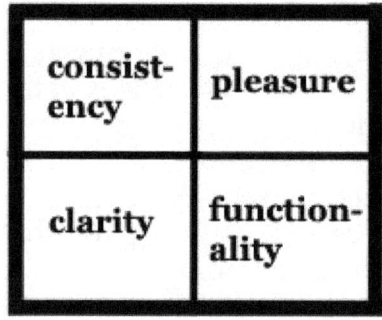

consist-ency	pleasure
clarity	function-ality

Diagram reads counter-clockwise from the upper-right.

Each stage represents a threshold of success.

Also called: (A) Motivation, (B) Work, (C) Emotion, and (D) Function.

Refutation of the Teleological View of Evolution - In Hull's book a distinction is made between teleological explanations and causal ones, a distinction which I think is somewhat too finely made. On the subject of philosophy and biology, it is inevitable to find some mergence between the theories of causality and those of teleology. While such combinations may not be practicable in a biology lab, it is unfair to say that an organism is nothing like an argument. Indeed, organisms, ontologically speaking, depend on arguments (of some form) for their existence. Without any capacity for argument, such as evidence of accomplishment or at least 'status', such an organism would not be alive. With the statement that organisms are like arguments, it may be extended that if an argument depends upon the conditions of an organism at a given time, then such arguments are reducible to values. If that is the case, then it can be said that if evolution amounts to organisms, and organisms amount to values, then evolution amounts to values. However, I argue, that given the variability of value-fulfillment, to a remarkable degree, there is no certain boundary between values and causal process. For example, an organism may have excellent adaptation, and poor tools, or excellent tools, and poor adaptation. Making matters worse, the two are judged to be evolutionarily equal. The semantic point is that the tools, however arbitrary, are relatively definitive for value. However, the tools are arbitrary. It is easy to see how some form of luck might influence one specific ape to use a stick in self defense, where another ape was unfortu-

nate enough not to use one. Although in
this example it is incidental (the second
ape might do better without the stick), it is
easy to see how in other scenarios the tool
makes the difference, independent of survival ability. Now consider this final point:
that language is such a tool which exists
with some remarkable degree of independence from evolution. Now consider Hull's
teleological view: given the synthetic nature of language, would we prefer being
someone who was poorly educated in a
primitive climate, with a profound grasp of
language, or being someone who is intellectually deficient, yet having some modern language ability, in the developed
modern world? According to Hull's view,
we must favor the primitive one, because
he has a better grasp of language, which is
a symptom of evolutionary advancement.
But this leads to a paradox: if everything
advanced is advanced relative to evolution,
then everyone must prefer being primitive,
since in the case of evolution, being advanced means being there first! There is a
great gap, then, between personal and social evolution which has little to do with
objective definitions of advancement. For,
if someone becomes extinct in a modern
society because of a lack of supreme advancement, where they could live a longer
life in a more primitive society with poorer
values, then they are supposed to prefer
the primitive society, but they do not! They
always prefer the more advanced society,
and yet this is precisely how they become

out-moded! On the contrary, how they achieve advancement is by being historically primitive. Amongst these contradictions, it can be stated that advancement is value-relative, and values exist almost arbitrarily. In other words, evolution is value evolution. This leads to a second paradox, which is that evolution is not defined as values, so evolution is not evolutionary after all! What a coincidence!

Reproduction - *Substantial Forms of Reproduction* -

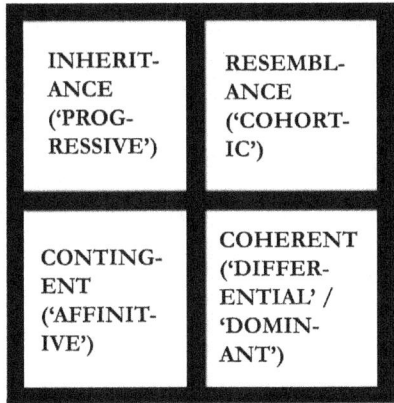

INHERIT-ANCE ('PROG-RESSIVE')	RESEMBL-ANCE ('COHORT-IC')
CONTING-ENT ('AFFINIT-IVE')	COHERENT ('DIFFER-ENTIAL' / 'DOMIN-ANT')

By this standard, conventional views of reproduction would be considered rather narrow. For example, sexual and asexual reproduction both occupy one category, 'inheritance'.

Resemblance is a type of reproduction that takes place when an organism coincidentally represents the properties of another organism. This can take place even when the two organisms are not directly related. If there is a causal link, however (or an inheritance link), it is easy to see how this mode might be primary, as may be the case with flowers selected for aesthetic appeal for instance.

Inheritance is the most widely understood form of reproduction, the passing on of actual coded structures (although it is also possible to imagine coded structures within the other modes).

Contingent or affinitive reproduction is when reproduction of a different species serves the purpose of one's own species. Therefore, such an organism reproduces to benefit not only its species, but also another. Thus, the reproduction of the first species is like (or benefits) the reproduction of the second species.

Coherent, differential, or dominant reproduction is also called 'reproduction that makes a difference'. This can occur within all of the other three types. For example, a flower that is especially beautiful can make all future flowers more appealing when it reproduces. A grandfather with dominant strong muscles can make his children all have strong muscles. A moss which softens the ground can make some humans develop more sensitivity and appreciation, and thus grow their culture. The moss gains a certain stature for that society. The same thing has been noted with bees for providing honey, with a caveat. Societies with the ability to gather honey become more cultured and stimulated, while those without the ability to gather honey lose those advantages. Thus, bees may have a two-pronged dominance (a created dependency) for a species that can gather honey. The reproduction of the species becomes the reproduction of a dominant product, a dominant activity, or some other form of dominance. But only within the fourth type of reproduction.

Resemblancy - enters as a response to the assumption of incoherent function. Instead of referring to some assumption of local laws, genuine functions (however specific, however complex) may be treated as derivatives of the most distant authentic law. Thus, the failure of comprehension is one of coherent interpretation (universal explanation) rather than being a matter of logical replacement. In this sense, however, the property of the function is not to physically reproduce the function but to synchronize with the general properties of nature. Hence, there is an analogy to the continuum of space and time, and the properties of quantum mechanics, suggesting that if resemblancy is not a universal rule, then it has at least become ubiquitous subjectively. Does this make it logically refutable? No, instead it entails complexity such as symbolics, with a high degree of contingency. Thus the concept of resemblancy becomes a useful razor for 'determining nature', in any case where it is decided that the specific must refer to the general, in essence, any case in which there is knowledge of biology. Alternate theories look very strange, but might be imagined in a world that is both contingent and dynamic, e.g. a world of pure perception in which interpretation is valued over truth. I suspect in such a case generalities will become primary, and perceptions will become authorities. I have used this analysis as a background concept for my notions of extraterrestrial variety.

Retroactive Retrograde - With the risk of defining biology solely in systems-terms, there is, in philosophy, a necessity to combine theory and experience. One of the problems that emerges is the retroactive sense of retrograde---the sense in which, although there may be many holes, systems are defined after the fact, by gradual appraisal of contemporaneous occurrences --- what may be called para-consistent data or properties. It may be said that it is the captivity of such data or properties that is actually para-consistent, leaving a lemma of the tangibility or else abstract definability of biology. Clearly there are several approaches: (1) Treat organisms as abstract data, which are open to interpretation, (2) Treat abstractions as

tangible elements, e.g. rational representations, or (3) Revert to direct sense-data in
determining the properties of biology. This
list of three may be considered a lemma to
attach to other aspects of biological study.

============[S]============

Schizoriginal Nomenclature - An emerging classification system that essentially leaves 'all doors open' to new organisms, and even new forms of classification. This system is more prevalent when bizarre exceptions are discovered, or when the specific origins of some forms of life cannot be known, or are known to be unrelated to the rest of life. Schizoriginal nomenclature is more likely with alien (Xenoid) species, and with the discovery of species that occupy bizarre habitats and special niches. Whether such nomenclature is even desirable is somewhat of an open question. Certainly there is some desire for coherence between organisms, although as the number of known species proliferates, this possibility seems less and less likely. The question is raised as to whether the true objective origin of species will ever be known. For, if even universes proliferate beyond our own, the source of a species may well be beyond our knowledge, or even recursive. This raises some potentiality for the value of philosophy in biology, and specifically dimensional forms of cataloguing organisms. However, whether the adopted system is schizoriginal is still up in the air. For now, it is a useful *erogative* for the acceptance and cataloguing of further new species.

Semantics of Biology - A cube demonstrating the types of exceptional conditions which are common in nature. An attempt is made to find numerous unique categories on this subject:

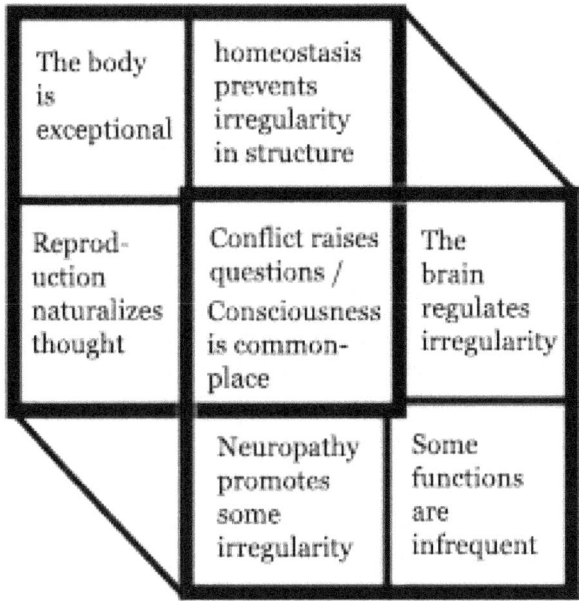

General deduction using a unified method:

[1/1:] Homeostasis prevents irregularity of structure, so that neuropathy promotes some irregularity, while the brain regulates irregularity, and reproduction neutralizes thought, so that the body is exceptional, and some functions are infrequent, while conflict raises questions, and consciousness is commonplace.

233

Sensuous Growth - is the trend, contrary to what might be called functional-computational biology, for organisms to develop a recursive, even self-conscious, sensory modality in which the sensations of experiences contribute to the formation of the parts of the body; More than mere responsiveness, what this implies is a reciprocity between inner and outer life, even events such as weather, temperature, or color may quickly change the body; This can be observed for example, in plants in the strong tapering of leaves in response to rain, and in humans with the growth of emotional response sectors in the brain, especially the forebrain, and enlargement of sexual organs or, in rare cases such as exceptional athletes, parts of the body related to specific tasks, such as feet for swimming or a nose for a semolier; In most, but not all cases, the designation will be genetic; But in some cases, conscious adaption leads to the development in the middle of life, as if by morphosis; However, sometimes these may be considered social (that is, invisible) or cognitive adaptations which have few outward characteristics; It is almost a greater puzzle to determine how this process may occur for viruses, if at all; Probably a dangerous virus develops a strong strategic relationship to an outside biological pattern, so that changes in the virus reflect the host before reflecting the virus itself, much in the same way as a parasite; There is thus potential dimension in comparing functional and dysfunctional parasites and functional and dysfunctional viruses, with further complexity when they are actually symbiotic with the body of the host.

Sickness - In humans is most commonly affiliated with allergies, and diseases with similar symptoms, such as flu and the common cold; These diseases only become life-threatening in cases of extreme mutation and high exposure, and (sometimes) to the very young and very old; The body system often responds to specific irritants such as dust, pollen, and infected materials, producing a histemene reaction, including the secretion of mucous and the urge to sneeze; Because *anti*-histemenes block the histemene process, they are classified as anti-allergic, but the effect may be to weaken the body to the actual pathogen, if it is serious; The body's natural defenses also include white blood cells, famously the target of Human Immuno-deficiency Virus (HIV); Ordinarily white blood cells chemically target suspicious viral material by releasing Killer-T cells, which destroy the particles; In the case of HIV, the white blood cells are attacked and decreased in number, so the Killer-T cells are not released; An initial response to basic and advanced illnesses involves improving defense, or eliminating offense; In the context of the body, this produces four basic categories:

Anti-Virus	Chemical Immunity
Acepsis	Boosted Immune System

In the case of a highly dangerous retrovirus, like Ebola, asepsis may eliminate some mutations, boosted immunity may prevent some infections, while anti-virus may weaken the remaining strains.

Simultaneity - Complex organ systems require simultaneous or continuous functioning much as butterflies rely on chance meetings in the wind. In the context of information that meets for the very forming of organisms, independent of any phenomenological latitude, a being's reliance on rules like simultaneity is highly pronounced. And, moreover, the exceptions to simultaneity are not always rules of nature. We would do better to duplicate the number of rules than to find exception to one rule. It may be noted that simultaneity is not always a symmetric principle. For example, the obverse-dependence of a single organism that later thrives upon a single food source is also a simultaneity between two life-forms, and such *relata* are not necessarily uni-directional:

Examples of Simultaneous Relata*:*
Consumption, Excretion, Desemination, Pollination
Parallelism, Survival, Utilization, Cultivation
Syngeny*, Symbiosis, Phylogenation, Parataxis
Display, Service, Scrimmage, Games.

*The development of two plant or animal species in the same environment, with or without symbiosis or threatening conflict.

Social Argocy - is the trend to look outside a group to solve a social problem; This may be explained by sexual biology or by a need to introduce or even delucidate (concoct) new information; Argocy has a reputation as a mode of arbitration, suggesting that it is concurrent with problems involving leadership roles, and indeed selection or selectivity such as that stemming from a search for social information is often cited as a source of complexity in leadership; Thus social argocy can be used as an explanation for individual leadership, an alternative to complexity stemming from the general inclination to simplify society, in response to an overwhelming *impression* of complexity; The determination of arbitrary (that is, enforced) selections by individuals thus corresponds to a social idiom of missing information and perhaps a search for social significance; See opposite at Individual Pluralism::

Species Fallacy - According to Godfrey-Smith:

> It would not by surprising if
> there were some human-
> specific genes that were also
> indispensible to living a hu-
> man-like life, but it would
> also be not that surprising if
> there were not...
> (Philosophy of Biology, p.
> 106).

What I think he is referring to is the common genes of an entire species. However, this highlights a common fallacy in biology, which is the mistake of comparing the entire species with the exceptional individuals within the species --- what I call the 'species fallacy.' However, Godfrey-Smith goes on to mention another interesting point , which is that "[T]he lifestyle of one species can be lived with the genome of another." In other words, although species-level definitions are uncertain at the individual level, entire species may in fact resemble one another in large ways. It is even possible (in theory) to have two species which are identical or nearly identical, developing in separate phyla. In other words, we should be cautious in applying the species fallacy to larger groups than individuals, or to do other than remark on the exceptional characteristics of a single species.

Species' Multiple Evolution -

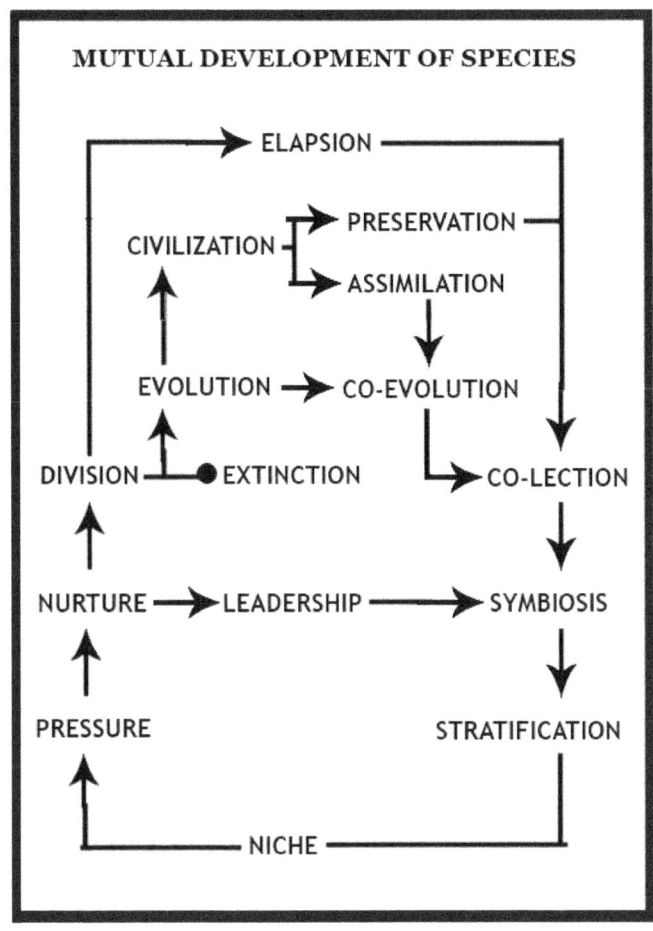

Subjectivity - While philosophically subjectivity is inherently formal or else meaningful or meaningless, and in psychology
subjectivity is accepted to be informal,
clinical, or irrational, in biology subjectivity genuinely appears, as the manifest difference between different types of entities;
If it is understood that 'type' is not a formality, nor merely informal in the sense
that it is embodied in real bodies---
systems with a life of their own---then it
can also be understood that there remains
some degree of objectivity: the actual natural resemblance or biological relationship
between entities; However, the nature of
this objectivity may remain obscure, even
though, critically, it can be supposed to be
manifest within experience, to the extent
to which perceptions are faithful to their
corresponding biological properties; It
may suffice to say that objectivity exists in
the abstract, in the same sense that it can
be proven empirically (strategized beyond
a doubt) that life forms are related,
whether through identical perception, sexual union, social space, or merely physical
dynamics.

Superficial Animal - One image/form
that may emerge in biology is that of the
superficial animal, what has become
known intellectually as *animalus terranus*
or ironically as 'the terrible monster'. This
encompasses figures such as a furry ani-

mal with many eyes, wearing many pairs of human glasses. The impression is of the hollowness of the animal identity, or the sense in which the animal must appeal to the human to maintain its dignity. We may ask, 'at what point does biology become superficial?' Is the difference only perspective? Is it possible to reduce sex, evolution, survival, to mere selfish political messages? At what point does science appear over-committed to the 'sham of materialism'? Perhaps real scientific achievements are not reducible, when they present real evidence of something new. But what is this but novelty? How absolutely can one make the claim that research is not itself reducible to sex, survival, and materialism? Perhaps, on the one hand, reduction is something which appeals to the scientist. But on the other, is the risk that any given thing is in some sense a monster --- a freakish animal, a coincidence of the properties of matter. At what point do the properties of the scientist have a sacredness beyond the contested territory of brute survival? The scientist, is, he or she admits, something of an animal. And perhaps it is only human success which makes lesser animals seem perverse. Maybe, as some have noted, without clothes, without a future, humanity can seem even more obstinate.

Survival Modes of Animals - Initial categories concerning general survival of animals:

select-ive	**domain**
dyn-amic	**respon-sive**

Selective domains have responsive dynamics;
Selective dynamics have responsive domains;

A second dimension of survival:

insignif-icance (recovery)	**force** (appear-ance)
intellig-ence (co-mmunity)	force (alleg-iance)

Force is insignificant when intelligence is allied;
The force is allegiance when there is

insignificant intelligence;

Translating the first two dimensions into a cube:

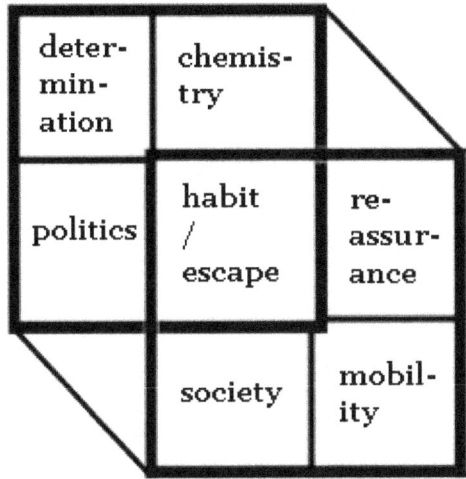

The chemistry of society is the habitual escape when the determination of mobility is the reassurance of politics;

Translating the eight boxes into four in order to introduce additional dimensions:

determination of mobility	chemistry of society
habitual escape	reassurance of politics

243

Adding a fourth quadra to the four-dimensional or higher structure:

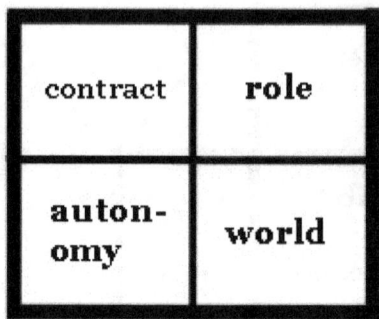

contract	**role**
auton-omy	**world**

The resulting eight categories are translated to result in the following cube, which is ostensibly a categorical hyper-cube, or tesseract:

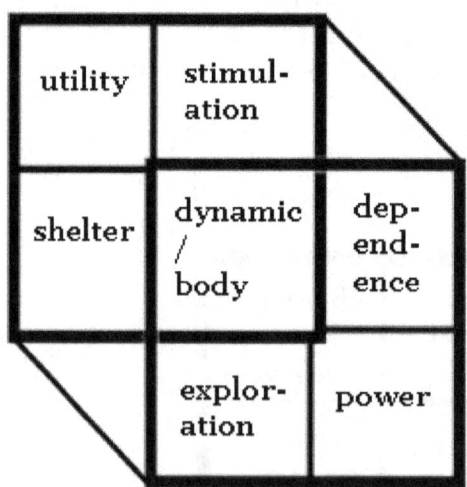

Stimulation-exploration is dependent on

shelter just as the utility of power is the dynamic of the body;

That's a good attempt at a five- to eight-dimensional survival structure which applies to a wide variety of animal phenotypes.

Symbiosis - May play an important role in organizing systems, particularly horizontally. Animals may interpret a mother figure in a wide variety of circumstances, particularly when the fight-or-flight cycle has come to a period of quiescence. Particularly through a mother figure, symbiosis plays the role of identifying the animal with the environment. Animals may undergo a period of trial in which it is determined if the environment is the right fit. If the environment is the right fit, a mammal will identify the environment with the mother. Other types of animals, such as sea-urchins, may identify the environment with the 'first explorer' --- which may be an organism of either gender that first tested the environment. Thus, the identification with the environment may be encoded genetically. Symbiosis is even more common between very genetically different organisms (say, a human and an aquarium) , than between similar organisms that may compete for food sources. However, social organization off-sets this. It may also be noted that organisms which are highly different from one another are not as likely to provide for one another's health, except via contractual obligations which have been realized gradually, perhaps over millions of years. So, in theory, it is primarily the identification with the environment that serves the function of symbiosis.

Systems Theory Biology

(Correspondence Theory of Truth) - The theory that organisms are only linked by processes, and only processes constitute entities; This theory stipulates that there is a functional component missing from unitary concepts of biology; Under this theory organisms may be of such types as a sexual partnership, or the breathing-oxygenation cycle, or the implosion of gas in the sun; Because of the ability to include energy states and not just organic states, this theory may be termed trans-material, and hence functionalistic. If there is a caveat about the properties of matter, it is thus likely to affect this theory, creating a material contingency similar to qua information or qua meaning, much in the same sense intended by the physicist Brian Greene ('black holes are the only material reality, all else is virtual') and the philosopher Frege ('genuine sense has connotation, not just denotation'). Connecting these independent statements to systems theory, correspondence is not just being dynamic, as is the convention of social science, but is also a function of complex information.

This should not defend any single thesis, except to create a functional dependency of dynamic upon information. Thus entities or organisms have taxalogic or 'global' significance in addition to having 'applied' or definitional value;. That contrast becomes important in parsing the ultimate connection between small isolated properties and assumptions about the overall functional role of the organism.

============**[T]** ============

Temporal Functioning (Lobes of the Brain) - Parts of the brain are 'rejected' by the natural brain process: this is not actual rejection, but a kind of locatedness, or logical subjection. The rear of the head becomes rearward not by an objective process, but rather by the almost-but-not-quite innate explanatory power of its *apparent* condition. This is no more apparent than in the case of a brain-damaged person who has exceptional gifts. Parts of the brain will develop without logical reference that occurs in normal cases. The result will be an un-locatedness, or sometimes hyper-locatedness of the mental process.

Terrete - (Not Turrette's)---meaning 'upon the earth' or 'game of return' or 'the thing in its place'---is the game of life, a fanciful expression that bears on biology in a number of contexts. These may include: (1) Paradigmatic sexuality and other forms of 'perpetual motion', (2) A landscape such as a mouse's maze, where success depends on exploration, (3) A context in which survival depends on one outward characteristic as opposed to adaptation. One example is an immortality drug, (4) Periods during which survival depends on one mode, such as wearing a space suit or learning to make tools. Although all these examples represent survival scenarios, it is easy to see that the cases also exist without risk, as variables of contingency, rather than annihilation. Assessing these four abstract variables still further, the context can be summarized as: (1-2): Population, and (3-4) Habitation, both under unconventional usage. If that is the case, there are further hidden levels of the game of life: (3-4-1-2) Habitation within a population, and (1-2-3 -4) Populating habits. This gives some access to the secrets of social science. Fundamentally, it is the variable that counts. But by no means does this mean that the variable cannot be generalistic, or function by reference to a specific pragmatic frame of reference. Indeed, general pragmatics serves to summarize the territory and potential of the game of life, opening biological frames of reference to the use of the most meaningful variables.

Theories of Incipient Biology (Meta-Theory) -

Good - 'God': spiritual reality.
    (1) Consciousness vs. Pain vs.
        Meaning
    (2) Art vs. Science vs. Experience
    (3) Life vs. Impossibility vs.
        Gestation

Evil - Nature: The natural limit-functions
    (1) Memory vs. Depression vs.
        Happiness
    (2) Temptation vs. Stupidity vs.
        Intelligence
    (3) Death vs. Tactlessness vs.
        Wisdom

Neutral - Principles: Things that depend on life
    (1) Systems vs. Unknowns vs.
        Knowledge
    (2) Processes vs. Waste vs.
        Efficiency
    (3) Values vs. Destruction vs.
        Construction

Any one of these juxtapositions could serve as a beginning point for life. These are almost like 'biological personalities'.

Thresholds of Consciousness - Life of all kinds seems to exist on a series of platforms or thresholds. At every point or shape in the structure, there are opportunities to adapt. Adaptation may depend on some of the most finagling details --- simple willingness, a desire for a certain type of food, or the appearance of domination over other species. That minor susceptibilities to adaptation make a difference is a thesis in itself, however, in this case, the important thing to notice is that these thresholds exist in a series, leading up to or beyond the condition of humanity. Imagine that the building one lives in is a kind of mask, and that one is more adapted the deeper one goes into the mask, and this gives us a picture of the nature of survival. The deeper---and more technical---we go, the better our platform for survival, the more tools we have, the more we depend on the infrastructure of nature or something beyond it. This has been described as 'the pecking order'. But the pecking order is not just top-down, it is also expressed as sheer functionality and sheer adaptation: essentially, the functionality of the platform.

Tractatus of Evolution -

1. At first, everything is energy.
2. Life, the first impulse, is a response to energy.
3. Very gradually, energy becomes associated with life.
4. Small bodies develop, to capture energy.
5. Then, feelers develop, to digest and use energy.
6. Products are guided through the body, creating vesicles, atriums, and cysts.
7. Chemical develop which serve the body, adapted to the Ph of the environment, and being tropophagic.
8. Brains develop as an extension of chemistry.
9. With increasing complexity, secondary

functions develop for the brain and other organs.

10. Complex functions become abbreviated into synergistic functions.

11. Synergistic functions become generic meta-functions.

12. At this point, development depends on specialized environments.

13. Functions now mirror the laws of the functional universe.

14. If it has not happened already, the orle for the organism becomes deeply meta-phorical, and then symbolic.

15. The function of the body then develops rapidly to meet up with the total potential significance.

16. If there is an opportunity, the organism becomes some kind of god---a highly functional organism, with any degree of compromise of reality.

17. Additional exceptions arrive, and the species is forced to adapt to new types of information. There is a risk of becoming archaic. Organisms are tested for their time-worthiness.

18. Those who survive the longest learn something, and those with the most experience learn something else.

19. Everything learns its place in the world. Everything bargains for satisfaction, as a function of (beauty, complexity, socialization, time, and distance).

20. Those who wish to know more, gain in complexity, and accommodate new unforeseen risks: an eternal way, a balance of difficulty and skill.

Tractatus of Genetics -

1. Genes happen for those that reproduce.
2. You can dominate the gene pool,
    but you become universal.
3. How about this: gender is not genetics!
4. But if it is, organisms tend to speciate.
5. Genes are as complicated as necessity
    plus complications.
6. That is an ideal state.
7. Conditions of nature prefer organisms
    that are hard-wired.
8. If not, organisms must be highly adaptive.
9. If adaptivity doesn't have sex appeal,
    organisms lose out.
10. Genetics is both conservative and
    risk-taking.
11. There is a priority on risk-assessment.
12. However, assuming one's children reproduce,
    reproduction tends to be worth it.
13. The first key is asexual.
14. The second key is mandated.
15. The third key is assessment.
16. The fourth key is advantage.
17. However, without conditions of nature,
    there may be no capability for survival.
18. Thus, the real key to genetics is surpassing
    context by integrating with the broad-
est
    context of environment.
19. In this sense, sex is perspective.
20. Knowledge without risk may be the central
tenet of genetics.

Transmigration, Diaspora, and Settlement - There is a clear relation between these types of trends, at least in human psychology. New homes may not be easy to adopt, but new generations acquire a sense of righteous dominion. This trend sometimes goes back many generations. Indeed, unless there is an encroaching sense of the weight of history, as occurred in Europe during the 1930's and 40's, little might get in the way of a society's sense of 'rightful desert' besides civil war, or other forms of defamation. Diaspora, which has been seen as an urban denomination of transmigration, occurs because of political or religious ostracization, not because of the more primal need to find food or a better climate. These three trends may be seen as 'triptych movements' which develop two genii to produce a third. Transmigration combines neediness with the ability to travel, to produce 'travel for neediness'. Diaspora combines political conflict with political drive to create relocation. Settlement combines resourcefulness with cheap land to produce new habitations. There is some question as to whether other major ideas, such as Panarchaism and Evolution could be considered triptych movements. Perhaps in a more abstract sense (formally rather than modally). For example, evolution could be a product of survival and reproduction. Panarchaism could be a product of time and nature. In any case, one may question the terms involved rhetorically, but not contextually. It is clear enough that in some cases those are the results that may be produced. And there is no question that those results are useful in many instances. However, it does do something to reveal the nature of exceptions upon reality. Clearly there is some relationship between one form and another, yielding intermediate types (perhaps this is similar to using the Semiotic Square, as differentiated from the Categorical Deduction).

Transverse Clausal Development - Is one degree less material than Abverse Clausal Development (the clausality affiliated with robots); Here functions are a result of organization such as government, personality, and competition---the human level----any examples of which serve as a pre-requisite for meaningful correspondence; Contrast with all the other clausal developments: Nominal, Universal, and Abversal.

Tree of Life [Dimensional Taxonomy]-

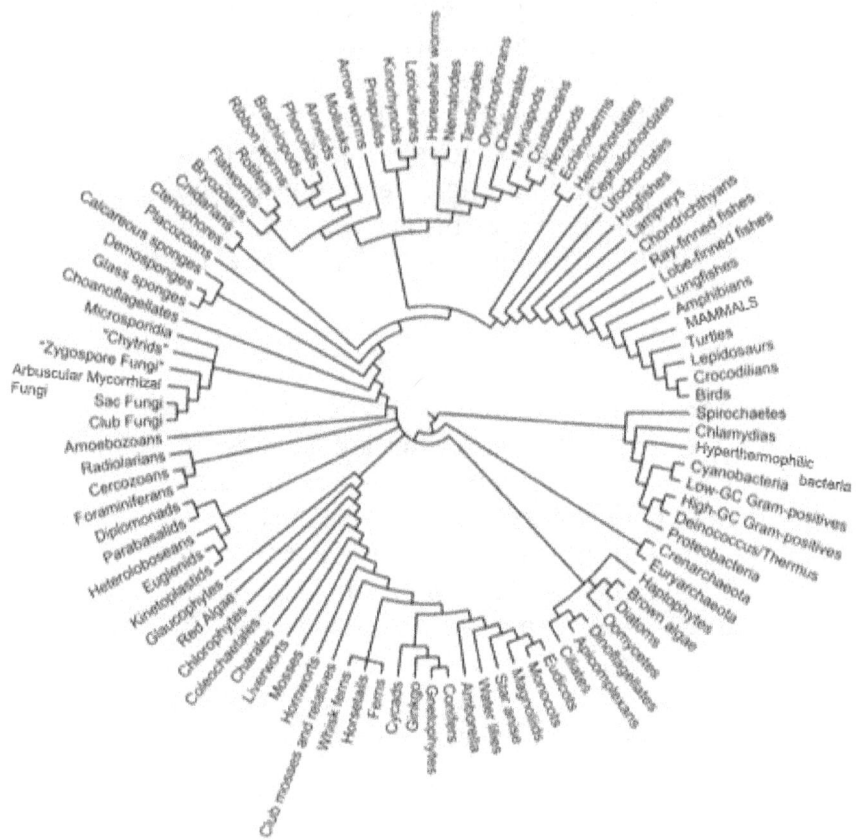

Notice mammals in the upper right. The roughly five major branches may refer to emerging fifth-dimensional consciousness.

Credit Due to Simplified Tree of Life / Phylogenetic Tree, © David Hillis, Derreck Zwickil and Robin Gutell, with permission via Drupal.org

Trees and Flowers -

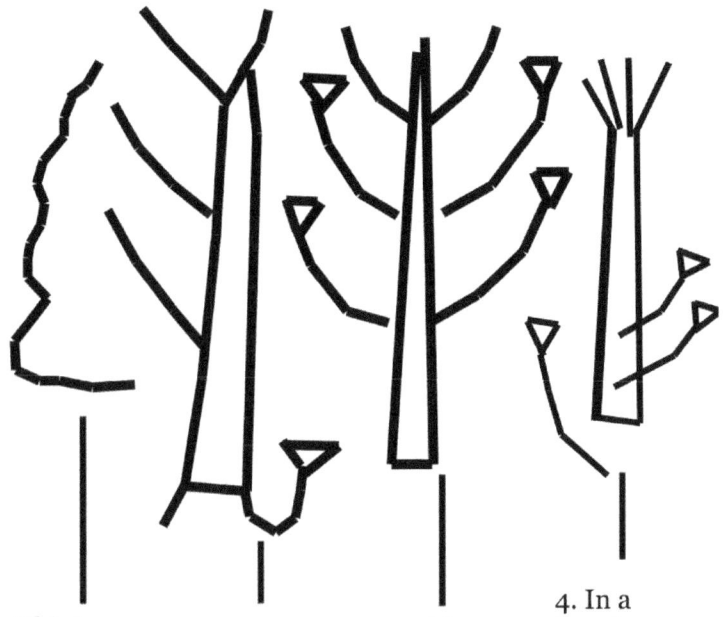

1. This type of tree appears as a kind of fruit, using macro-repr-oduction (cumulative spores)

2. This type of tree has a root-network that interacts with the pollen from the flowers

3. This type of tree is instead a flowering tree, using flowers for it's own reproduction

4. In a fourth type, the tendency is for scattering, where combinations are happen-stantial.

One can guess that this series runs roughly from least to most threatened biome.

Nathan Coppedge

Tropics and Distropics - In the most
modernized feel of things, biology can be
conceived as relating to entire fields of development, including all the conceptual
byproducts of human behavior. Each concept, is, so to speak, a 'tropic' of future biological potential. Byproducts of species can
even effect other species, much as one species of creeping vine learned to grow flowers. Distropy on the other hand, occurs
when one species paradigm overwhelms
the species, and moves on to another species. This is like abstract evolution. Both
tropics and distropics are useful tools for
studying the future of biological development, as doubtless new paradigms will
arise, and old ones will be overwhelmed .

============**[U]**============

Uber-Adaption -

Four forms of adaptivity.

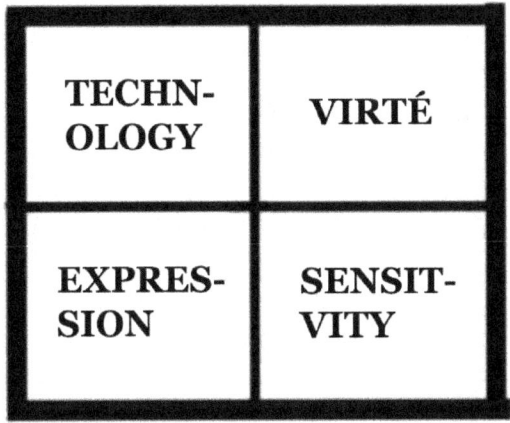

TECHN-OLOGY	**VIRTÉ**
EXPRES-SION	**SENSIT-VITY**

The term virté provides a segue to uber-formulas of adaptivity. Virté is, so to speak, strong virtue, such as vitality and paradigmatic survival. It is a term weighted to mean fluent, easy accomplishment which, if taken to a greater extreme, could connote bad behavior. So it also connotes efficiency. Using virté as the first category, one needs an analogy to permanent development. This is summed up in the word 'technology', a word which in philosophy has a similar meaning to adaptation. What is missing from these prior examples? Expression and sensitivity to render the context dynamic and conscious. By opposing or melding with sensitivity, technology provides an infinite platform. Virté develops from the subliminal as an equation with expression connoting pragmatics, allowing a context for sensitivity and technological virtues.

Unessential Compounds - Considering the level of human adaptivity,
It is not surprising that there are very few compounds to which the body does not have some accustomed response. Carcinogenics---compounds which cause cancer--- are relatively rare, and tend to be in some sense 'more dangerous than dirt'. Asbestos, for instance, was only introduced as an industrial element recently, and thus, there is a lingering possibility that there simply was not yet time to adapt. On the one hand there are 'positives' of this, such as perhaps there will be improved trends of mutation in future generations, while on the other hand there are negatives----for there is no denying that some substances are waste products, for example, things high in Alkali or containing few if any valuable fibers or minerals.

Universal Clausal Development - In contrast to Transverse Clausal Development, assumes that one process of organization is the development function. In this way it can allocate that every process has an implicit reference, and hence genuinely organized function. Non-universal approaches do not feel that real organization is possible except when there is direct influence on information---a Nominal Clausal Development view. The other two forms totally ignore the prospect of the universal individual, by accepting universal correspondence (the Abverse view) or coherence (the Transverse view).

Universal Flesh - May be seen as one of the things that organisms grapple with, within their own echelon of development. This implies that the substrate of development is not always friendly change. For example, clams and other mollusks have embraced an entire covering, implicating a certain degree of sacrifice, whereas people find comfort and attraction to the softness of the skin, creating vulnerability. The trend has roots in fungal growth (which has at any rate been imbibed by the animal when these substances are eaten), where the confusion between mouth and sexual parts can be observed in the arbitrary or environmentally determined reproduction of stamens such as in the case of moss, whose primary function is numbers, hence envelopment, and penetration is replaced merely with growth, or perhaps growth and conquest against other species. The incident of fungal growth may be compared to other incidents that we term grotesque, such as when too much skin grows over the mouth (when such is not eliminated by cutting) or when plant skin is used to replace human skin in an experiment. These examples have some resemblance to cancers, but illustrate that it is no more than optimization which makes the distinction.

Universal Principles of Biology -

Universal properties were not made this clear with Darwin~~~.

1. A daisy-like flower with drooping leaves will develop four stamens positioned far from the center of the flower. Flowers that are not this flower tend to be seductive (many stamens, or much pollen), or have ungainly growth (like lilacs), or else develop elaborate methods of survival (like orchids).

2. Four people tend to develop two children, so that two children can meet two children. Similarly, eight creates four, sixteen creates eight, thirty-two creates sixteen, etc. (these are matable sets: some parents may eventually produce more, but only by competing).

3. A tree develops one branch for every aspiration: (A.) a tree with many branches in the trunk aspires to have many leaves. (B.) A tree with cantilevered branches tends to have no leaves, and instead bristles, etc. (C) A hypothetical tree that sprouts from a

bristle tends to repeat branches in a zig-zag pattern. (D) A hypothetical tree that sprouts from a leaf tends to have some of the properties of the branching of conifers, such as spreading horizontally or molting.

4. Animals are aggregates of objective properties only modified by exceptional environments (things that crawl tend to have tails, just as things that move quickly tend to have defenses). For if an animal does not have an appropriate environment, it does not survive. Some exceptions are adapting to change the environment, and universal organisms (a further suggestion is an adaptive or universal habitat). Plants are modified by local processes the way animals respond to their own ability to adapt.

5. The mark it makes is the real memory of a thing: bears are claws the way humans are writing, noise, or urban sprawl. A plant's memory is internal to those that eat it: poisonous or fecund. Processes are where chemistry occurs, whether internal or external. At some point, measurable from the recent present, individual organisms will begin to make evolutionary leaps based exclusively on signs. If signs infinitely inflate, so too does the future of the biological signifier.

Unpredictable Exception - Biology is different from philosophy in that exceptions depend on empirical experience rather than a conceptual system. The exceptions may depend on matter more than logic. Where philosophy is contingent-to-contingent, biology is more likely to be meta-contingent: since no original contrast is created except in experience, biology has no choice left but to quantify a singular or compound property. The context of physical exceptions is often what is called quantum realism. Although there may be no evidence that a given property has changed, science knows that change is more likely without observation---part of the growing science of the macro-quantum, which may lead to new theories of organic chemistry. Thus, biology, if it is informed by physics, is more prone to hidden plethoras, and advantageous simulations. The behavior of an organism can be interpreted as a kind of unexpressive puzzle, as a practical coincidence (e.g. always expressing subtle unknowns), or as a simulation of even deeper significances (e.g. cosmic facts) which go unobserved. This final factor is what may be called divine biology.

Urban Onion - Everything alive is, in some sense, emergent. Everything is lumped together like an onion, an urban onion. Part of this is that everything takes some of the same variables (heat, cold, dust, a view of outer space, a level of entanglement). From the enmattered, ersatz conditioning, the new or old conditionality of entities is made. Chemical formulas of the greatest usefulness are born. The language-qua-entity---the alignment---of organizations of materials is born on the basis of politics and longevity. But further factors---neutrals---are in play. Conditionalities not only of conflict, superiority, health, and intelligence, but also integration, adaptation, survival, and procreation. It is these additional factors which characterize the urban onion. Factors of dint and near misses. Factors that characterize individuals as much as they characterize society. Variables that integrate between the ecosystem and the individual of the species. It is then for the urban or ecological concept to define the boundary between top-down, inward boundaries, and bottom-up, outward boundaries. By defining the two things, borrowing Stephen Gould's phrase, the urban onion becomes a mutually-interpenetrating majesterium.

========[V] ========

Viable Epochranysm - Ignoring habits as non-exclusive, and time as inconclusive for exceptional cases, there appear to be cataclysmic or para-archic events at many scales which influence not only development, but the largest impressions formed by species about their environment. What I mean is that accepting that some environments are arbitrary or determined by the organism, it appears that the 'objects in play' effect natural development with no less force than the genome itself. This becomes a link to the subtle interplay between small amounts of even taxonomically unrelated genetic material. Although the genome is coincident with original potential (that is, species-level choices), it is likely that the genome is not the most original element (being a function), unless the genome participates directly in decision-making cognitively. It is more likely that the interface with the world, or behavioral necessity, is a greater factor in combining with the original impulse of self-nature. This would go a long way towards explaining biological dysfunction in isolated environments, perhaps providing a corollary to experience. The individual, whether he or she may or may not be superior to genetic elements, certainly has an original impetus to commit to a structural liability like genetics. Otherwise, it cannot be said that individual existence has the same justice which biologists attribute to it ---the individual might be seen as 'more mad than the system' when the individual has already been called nothing more than a system in itself. So, (1) The organism's liability is original, (2) The liability is largely an ability to understand the world, however, (3) The world is automatically larger than the reasonable context of development, considered as characteristic, therefore, (4) The individual, having experiences of the world, must be larger than his or her genome. The implication is that the genome is a function between the self and the world, which may be more technically complex than the person and simultaneously less original than the environment (e.g. citing 'use of symbols'), or otherwise the environment may appear simple and the individual may appear original (citing the concept of original self). But the only exception to the second case appears to be a genetics which responds to the environment. The two cases may also be mutually-inclusive. I will consider an exception to the entire case of object-oriented biology, and it is the case of qua-formal influences upon the society, group, or individual. These are things like fears, beliefs, and assertions which play an intermediate role between complexity and originality. Essentially, although formality may take place through a contractual role, there is no reason to believe any form of contract is binding except through artificial means such as extenuating circumstances. Barring extenuating circumstances, it is easy to defend the second point, that individual are original. So the choice is essentially between creative genetics and creative identity. This raises a series of questions, which are also serve as a miniature tractate of influences upon genetics: (1) Is nature extenuating?, (2) Questions of economy and imagination, (3) Is complexity original?, (4) Is the genome artificial, e.g. 'reflective'? What constitutes knowledge genetically? Would such knowledge be beyond genetics?, (5) Are there unnecessary constraints on biology which exist for artificial reasons? The general importance of this theory, however, is that changes may occur because of features of the larger built environment, which influence the permanent character of species only through extenuation and therefore permanence. So, in this sense, the genetic structure is an unfortunate coincidence of extenuation, which only incidentally expresses a structure of material, non-psychological development. In this sense also, psychology is primary to biology, as an explanation for creativity or else 'vertical' entification. However, biologists have already answered that politics is in some ways more important, raising a triplicate problem of how to resolve politics, psychology, and value simultaneously. The clear answer here is of a number of kinds: (1) Biology constitutes knowledge or experience of a number of kinds, corresponding to the types of organs, functions, and faculties of which life consists. This is an egalitarian view which recognizes that some political solutions are not equitable solutions. (2) Biology is primarily a macro-system which obeys physical rules about structural functions. These structural functions determine that objectively, radically, often permanently, specific structural variations favor specific structural evolutions. In this view, survival is expressed as radical differences, and life follows tough-as-would-have-it rules which determine however simply but also *exceptionally*, that one organism is favored over another. Real differences are often real successes or real failures, equatable to moral differences. (3) Another, final view is more pessimistic, but also more intriguing than the others. It is the view that optimal functioning is always an exception to the rule, and never has anything inherently to do with the structural properties of the organism. In this view, the few politicians and power-brokers of the animal kingdom always make away with all the prizes, regardless of the level of evolutionary development, and regardless of the apparent balances of power. In this view, the most surprising developments occur when the most functional organisms are not the most political organisms, which could occur for a wide variety of reasons. It is also the case that animals of all forms are beholden to nature. There is nothing about natural evolution that says that extinction is impossible. In this view, it is politics, not functionality, which is the basis for conscious existence.

There is nothing about functionality which makes someone more or less miserable. Life can be significant even if it is dysfunctional or doomed. The other two choices provide alternatives: the systems-functional view, which may incorporate networks of organisms, and the egalitarian view, which is entirely lopsided in favor of exceptional functions, independent of sacrifice.

Viruses - Have more in common with bacteria than with humans. In a second sense, their threat requires reproduction. In a first sense, their threat requires chemical sophistication. Viruses can be described as typologically two-dimensional, because of this dual dependence on reproduction and sophistication. Nonetheless, there are many viruses such as the one that collects in the dander of cats that have altered human history, sometimes through subtle effects on psychology. In the sense of reproduction, the virus tends to be unthreatening if it has some compatibility with the host organism. In the sense of chemical sophistication, the complexity of the virus may prevent it from reproducing, unless it fulfills a niche. The most sophisticated viruses are often parasitic or phagocytic, because this allows them to reproduce within a living or dead body with less hindrance from the immune system. Another major type of virus is one which depends on feces or some other external substrate to reproduce. These types may rely on populating within the substance, and then wait for the substance to come in contact with a non-infected organism before beginning their primary life-cycle. Because of their bizarre chemical state-of-affairs, viruses usually rely on host organisms instead of food, limiting their capacity to procreate and dominate.

Volitional Mechanical Guide to Biological Life - The theory of perpetual energy called volitional mechanics has some things to say about fundamental principles of biology. Surely, the forms of mechanics that could undo the First Law of Thermodynamics [See my Nov. 10, 2013 successful over-unity experiment] have a lot to say about the qualifications of biological formulae.

The First Principle: Continuous Motion: Animals and plants must continuously work for survival, or some better principle. They must avoid risks.

The Second Principle: Unity
What allows an organism to survive is the ability to meet basic criteria. Beings that do not have necessary survival have statistical survival.

The Third Principle: Over-Unity
A combination of factors make some organisms more likely to achieve immor-

tality.

The Fourth Principle: Volitional Energy. A large number of moving parts synchronized with time is what determines repeatable energy.

The Fifth Principle: Volitional Equilibrium. Sub-systems are inefficient, and quantity is valued.

The Sixth Principle: Volitional Efficiency. Equilibrized energy produces values for quantity and energy against time and sub-systems.

Other principles: (A) The number of advantages may have to exceed the number of temporal dimensions. The easiest way to do this is informationally. (B) It is better to get things done sooner than later: time-consciousness. However, patience is valuable when time serves a purpose. (C) Lightweight parts increase the value of masses used for any function. Therefore, there is a mass-function equation that favors function over mass. (D) Cheating is permissible when it involves exceptions rather than rules. Sometimes the property of one thing will contribute to alleviating another property following a secondary rule. (E) Finally, in general, the difference is if something is 'in the clear': what does it achieve? Then, you can ask what it is.

============[W]============

Windfall Physics - Some properties sometimes develop which seem especially paradigmatic, which appear to break all the ordinary rules of behavior. And while these characteristics may not always be as useful as they appear, it seems true that the properties of these characteristics emerge solely through some extra dimension of reality, whether it is material complexity, evolution, or additional sources of energy. Properties such as (prior-embodied paradigms, genetic stimulus, and social pressure) combine with properties such as (intelligence or leisure, labor or duty, and circumstances or adaptivity) to reach the new properties of bodily or mental characteristics. These characteristics develop primarily in the order of commitment → mentality → appearance. Properties such as new appearance and new chemistry tend to develop late. Properties such as new functions and modifications tend to develop at an intermediate stage. Properties such as motivation and a chance at survival tend to develop at an early stage. Exponential factors include environmental factors and intelligence. It can be seen, for example, that dimension proceeds in stages: (A) Response-environment, (B) Exceptional response environment, (C) Standardization of exceptional responses, and (D) Exceptional environment. Further stages are possible, al-

273

though they tend to realize combinations of the previous levels with greater exaggeration. Here are some examples of organisms which fit the four levels:

(A) Herd animals, sea sponges, worms, ants, snakes
(B) Chameleon, Human, Fire-ants, Sea-snake
(C) Chameleon-snake, Psychic person, Venus fly trap
(D) Hive-mind, Conscious A.I, Immortalized being

An application of biological deduction to these levels yields the following assessment:

[1. Youth. 2. Maturity. 3. Death or Immortality]

I. Title: 2
1A: B : 3C: D
I. Maturity
1. Youth is the response-environment so that there is an exceptional response
2. Standardization occurs through death or immortality so that there is an exceptional environment.

II. Title: 1
3A : B : 2C : D
II. Youth
1. Death or immortality is the response-environment so that there is an exceptional response.
2. Maturity occurs through standardization so that there is an exceptional environment.

III. Title: 3
2A : B : 1C : D
III. Death or Immortality
1. Maturity is the response-environment so that there is an exceptional response.
2. Youth occurs through standardization so that there is an exceptional environment.

============[**X**]============

Xoration - ("Process of the State of Change") Elsewhere I have described how in biology there is a relation between systems and consciousness. Systems are the cullination of consciousness. There is, however, at least one more principle, which is Xoration. Xoration (named after 'inexorable'), is the inexorable 'thisness,' the insatiable *haecceity* of experience--- it is what connects youth with adulthood, learning with mastery, mystery with purpose, and permanence with variation --- it is the stuff of life, interpret it as you will. With *xoration* it is possible to apply logic to biology. '*Xora*' becomes the variable which has equivalence with a successful process such as maturation, intelligence, or survival. With '*Xora*'-equivalence, survival is an under-recognized accomplishment. Xoration can be used to express shortcuts or special genius on the subject of survival. Unexpected successes are '*Xorei*'. Pragmatic approaches treated as variables are '*Xoreis*'. The overall Xore is the Xora, the sense of universal participation with nature. Factors like urination and bowel movements, sex, eating, and exercise can be basic forms of *Xorei*. Conversation (such as gossip and negotiation), writing, mathematics, art or symbolism, and other forms of cognition such as science or psychology may be seen as higher forms of *Xorei*, which map human social and value concepts. The Xorei project may be to link species traits with more global traits, such as by interpreting them as expressions of more primitive principles, or by linking them to psychological traits such as confidence, rule-making, and rule-breaking. *Xoration*, as an expression of consciousness and systems, becomes a vitally civilized answer to biology, one that encompasses such notions as modular value and intelligence networking, not just as teaching concepts, but as political and ideological ones, with roots in the most fundamental learning processes. Through Xoration, it may be possible to discover the roots of human learning, the universal impulses shared between alien species, and even the hidden meanings (ironies, epiphanies) of biological structures.

END OF THE MAIN SECTION

Nathan Coppedge

APPENDIXES

Nathan Coppedge

APPENDIX (I).
Animal Phenomenology

I. Appearance and Defense
1. Human, which smells badly and makes weapons.
2. Neon fish, which looks poisonous.
3. Jellyfish, which is beautiful and deadly.
4. Salamander, which loses its tail.

II. Energy and Reproduction
1. Mollusks, which are low-energy filter-feeders.
2. Cows, which must continually eat grass and are mated by husbandry.
3. Macaws, which have elaborate mating rituals and seem indifferent to human presence.
4. Caterpillars, which mate and eat during a certain season.

III. Survival and Dependence
1. A lion that eats its young when faced with competition.
2. A beetle that finds a particular artificial image more attractive than it's natural mate.
3. A hummingbird eating insects out of a piece of fruit.
4. A dung beetle that eats dung.

IV. Adaptation and Extremism
1. Penguins that adapt to be detectable to their mates in snow and ice.
2. Cheetahs that must catch prey
3. Helophiles that live in continental faults beneath the ocean.
4. Dog breeds that appear a certain way only because of selective breeding.

APPENDIX (II).

DIMENSIONAL PROPERTIES OF PLANTS & ANIMALS

Many plants can regenerate from the dead.

"That which consumes itself consumes anything" ---Aristotle

Many animals can survive without sunlight.

Perpetuity might eventually depend on adopting the properties of other species.

APPENDIX (III).
BEAUTIFIC VIEW OF ORGANISMS

Metaphysical Beetle (black-shelled beetles),
Source: Gestalt

Universal Grasshopper (yellow grasshopper),
Source: Gestalt

Archids (metaphorical animals),
Source: Logos

Man-Size Mantis, Source: Logos

Pod Turtles, Source: Logic/Logos

Artists, Source: Platonic

Gemius (living gem), Source: Analysis

Bioclave (living instrument),
Source: Analysis

Divinus Relativum, Source: Dimension

APPENDIX (IV).
QUALITATIVE TYPOLOGIES

ANIMAL TYPOLOGY

ETHICS

BODY   METABOLISM

NERVOUS REPRODUCTION DIGESTION

THOUGHT COMMUNICATION IMITATION EXCRETION

KNOWLEDGE AFFECTION   FEAR   DISGUST   CONFUSION

PRESTIGE IMPORTANCE USEFULNESS OBSESSION INDEPENDENCE SEPARATION

PLANT TYPOLOGY

SYNTHESIS

CONSUMPTION   REPRODUCTION

RESISTANCE   COMMON-NESS   VARIATION

TYPOLOGY OF SPECIES SELECTION

INTELLIGENCE

COMMON
ENVIRONMENT   INTERACTION

GET   GOT   GRAB
(NECESSITIES)   (RESOURCES)   (LUXURIES)

TYPOLOGY OF THE ORIGINS OF ACTION IN ANIMALS

[1-D: ATHLETE]: FITNESS

[2-D: ACTOR]: SICKNESS, ACTION

[3-D: PLAYER-KILLER]: SELECTION, NUANCE, AGGRESSION

[4-D: PLAYER]: INTELLIGENCE, LOYALTY, STRATEGY, HAECCEITY

[5-D: LEADER]: LOGIC, LAW, ORDER, FULFILLMENT, PROMISE

[6-D: THINKER]: HABIT, TECHNIQUE, PARADIGM, NEOLOGIC, THOUGHT, DESIRE

[7-D: GOD]: PAST, PRESENT, FUTURE, TIME, WORLD, THE PSYCHIC, SOUL

APPENDIX (V).
TYPOLOGICAL
ORGANIZATION

(ALL-INCLUSIVE LEVELS)

UNIVERSES
/\
PLANET SYSTEMS
/\
PLANETS
/\
ECOSYSTEMS
/\
SOCIETIES
/\
ORGANISMS
/\
ORGAN SYSTEMS
/\
ORGANS
/\
TISSUES
/\
CELLS
/\
ORGANELLES
/\
MOLECULES
/\
ATOMS
/\
SUB-ATOMIC PARTICLES

APPENDIX (VI).

DIMENSIONAL SIGN. OF MINERALS TO ORGANISMS

MAJOR MINERALS
CALCIUM → Systems
CARBON → Life
CHLORINE → Water Processing
MAGNESIUM → Molecular Compounds
PHOSPHOROUS → Ripening
POTASSIUM → Emotion
SODIUM → Processing
SULFUR → Metaphysics

MINOR MINERALS
COBALT → Generation
COPPER → Change
FLUORINE → Translation
            Not Communication
IODINE → Metabolizing Illness
IRON → Seriousness
MANGANESE → Synchronization
MOLYBDENUM → Input
SELENIUM → Groundedness
ZINC → Transfer

APPENDIX (VII).

BIOLOGIST'S GOLD

Somehow I don't mind fruit flies as much if they're in the sink.

What does this mean? One can't know. Or else, many things.

In advanced human logic, primitive animals eat with their ass.

By this logic (this oppositeness of mouth and anus), mammals don't celebrate their consciousness through performance-survival, but rather through boring, cumulative activities.

There is a tendency that excitement for humans is ironic, not basic.

1. Microscopic
Suspensions.

2. Insects
Leaks and spills.

3. Mammals
Reaching and breaking.

4. Humans
Mining and building and growing.

APPENDIX (VIII).

Eucaleh Terrapin on Biology

"Perhaps there were degenerate forms of Tyranosaurs"

"Living the life of one thing isn't living every thing"

"Step in a puddle and you might step on a mud turtle"

"Take a tapered wing into the air and it sings"

"To the common child, a piece of candy looks like a modus operandi"

"Adults don't find traction without a means of satisfaction"

APPENDIX (IX).

Delineation of Consciousness

Idea
↓
Inheritance
↓
Memory Experiences
↓
Consciousness/
Behavioral Modification
↓
Psychic Experiences
↓
Time
↓
Natural Legacy

APPENDIX (X).

NEW VOCABULARY

Tropics: Conceptual potential.
Distropics: Conceptual migration.

Meta-Biology → Metabolics
Entics → Interbolics
Time Characteristics →Modular Phenome

Epochranysm: harsh adaptation, not extinction.

Manifold: Dimensional

Nathan Coppedge

APPENDIX (XI).
AXIOMS / EXPRESSIONS:

Law of the Scientist: Not everyone was set up with narcotics, but those who weren't were better off (usually)...

Criterion of Planets: The tendency will gradually be to recover from objective disasters, and then to reproduce the characteristics present on planets where those harmful conditions do not exist.

The general trend of evolution is something like the inversion of the least functional laws of nature.

Specific structural variations favor specific structural evolutions.

The definition of basic function shows the limit of complexity in all types of organisms; the psychological parallel to this is language.

Infinite reaction versus permanent process is the metaphysical condition of plants and animals.

Quantity versus scale is the property of confrontation.

An organism's 'idea' is greater than or equal to the idea's recursion in the case of successful adaptation.

To have a different reality is to have a different idea *ab origine* (such as inheritance).

Psychic experience is what impels animals to have futures.

The legacy of an animal is what resolves from the collapse of its fifth dimension.

APPENDIX (XII).

PRINCIP-PUZZLE

Our bodies are proportional to what we eat not only because of the energy content, but because digestion requires that we turn sideways every 24 hours, or we get indigestion such as choking symptoms. Thus we 'learn to live within a day' which also has spatial ratios. E.g. movement is more laborious for larger beings, and so, our day seems shorter relative to indigestion if we are larger relative to other creatures. This temporal dependence may occur in other animals or cells, even plants, e.g. the dependence on periods of darkness to process sunlight and maintain homeostasis.

APPENDIX (XIII).

PRE AND POST HUMANITY

Coincidae	-	Causes
Virae	-	Stories
Parasitae	-	Legends
Algae / Protobiotics	-	Myths
Worms	-	Agendas
Mice	-	Plans
Herbiverous Mammals	-	Societies
Predatorial Mammals	-	Legalities
Apes	-	Programs
Humans	-	Ethics
Post-Humanity 1	-	Adequation
Post-Humanity 2	-	Formility
Post-Humanity 3	-	Idyllization
Post-Humanity 4	-	Virtexis

[Adequation: Radical Pragmatism]
[Formility: Type Realizations]
[Idyllization: Progress Adventure]
[Virtexis: Objective Re-Definition]

APPENDIX (XIV).

CRITIQUE OF BIOLOGY

[BIOLOGICAL PARADOXES]

1. Life is completely open-ended, and yet species are seen as having functional roles in their environments. Either organisms can be described by ulteriors, as if they do not have functions, or they have functions which are self-generating. In either case, nature is not functional except via some form of description. There is thus no clean, causal sense of definition for biological function, or otherwise, biology is reducible to form. If it is reducible to form, philosophy says, then it cannot be reduced to causality.

2. Psychological identity is not a brutal survivor unless it is playing a dubious game for exceptional preference. This is not according to Kantianism, but instead, the view that to have preference (to have a good mind), already requires preference. If there is a struggle, it is dubious, because the gains have not yet been acquired. Apparently, only animals which have already been selected can be selected. Otherwise, survival is psychological and potentially arbitrary, or may be acquired by those who are undeserving (vis. having bad experiences).

3. The functional general ecosystem is not

necessarily the ecosystem that puts itself to
a test. There is no notion that skill is what
is at stake in the survival of global speciation. If risks were taken in the survival of
species in general, losses would be taken.
If the entire program is not a sham, then
there is some degree of artificiality built
into species survival. If there is not artificiality built into the system, then what is
suggested is that multiple species have
survival potential, or the entire system is
doomed.

4. While the position of someone's arm is
vastly contextual, independent of context
the arm has little inherent survival value.
On the other hand, the function of the arm
is universal, and always has a functional
relationship, with or without a context.
The success of functions is the success of
context, while the success of a given state-
of-matter is merely its function. Apparently, the reasoning for functional biology
is circular! Hypothetically, there could be
immaterial functions, they just wouldn't
have the context of material functions. It
might be more practical to compare
heaven to earth than an arm to the function of an arm! It's as if *relata* aren't related. The only calculations are functions
that happen to exist when categories
merge::

WEB LINKS

Article Directory (Genetics, Microbiology):
http://www.ncbi.nlm.nih.gov/pubmed/

Online Biology Tractatus
http://taxonomicon.taxonomy.nl/TaxonTree.aspx

Interesting Diagrams [Materialist Perspective]
http://www.engenetics.net/paper/diagrams.php

How Humans Might Still Be Evolving:
http://finance.yahoo.com/news/human-face-might-look-100-171207969.html

Modular Brain: http://www.ncbi.nlm.nih.gov/pmc/articles/PMC2784301/

BIBLIOGRAPHY

General Zoology, Olson et al.

How to Create a Mind. Kurzweil.

H.R. Giger's Biomechanics, Giger

Malay Archipelago, Alfred Wallace

Moral Machines, Wallach and Allen

Origin of the Species, Darwin

Philosophy of Biological Science, Hull

Philosophy of Biology, Godfrey-Smith

Physics of the Future, Kaku

ARTICLES CITED

"Epistemological Issues in Omics and High-Dimensional Biology: Give the People What They Want". Tapan S. Mehta1, Stanislav O. Zakharkin1, Gary L. Gadbury2, and David B. Allison; *Physiological Genomics*, January 2007, Vol 28. p. 24-32

Nathan Coppedge

INDEX

INDEX
(ALL CONTENTS SOURCE ALPHABETICALLY)

> →: See only under the following
>
> **See Also:**
> At category heading, see category within In-
> dex
> At subject description, see primary contents
> of the book

ADAPTATION
    Adaptive Thesis (→Categorical
        Adaptation, *N-Modifier* →
        Genetics)
    And exceptional biology (→Deferential
        Exception)
    and miscellaneous typing (→
        Categorical Genotyping)
    Categorical Adaptation
    chemical (→Barr Body)
    contingent adaptation (*game of life*→Terrete)
    Exceptional Biology
    general (→Cube of Life)
    fillistary roots (→Fresh Air and Sunlight)
    infant adaptation (→Paradox of Prematurity)
    Leadership (→Social Argocy
            →Fresh Air and Sunlight)
    mal-adaption to drugs (→Psychic Addictions)
    Parasites and (*Negative Ingestion*→Ingestion)
    similar meaning of technology (→Uber-Adaption)
    Uber-Adaption
    uncoded possibility (*Negative Ingestion*→

Ingestion)
useful for modality (→Categorical Genotyping)

ANIMALS (See also under Body, Organs, etc.)
    Animals Profile
    Butterflies
    Monkeys
    Paradigmatic Animals - Four Types

BEHAVIORS
    Aggression -
        Biological Conflict
        Praetorialism
    Breathing -
        Isomorphic Breathing
        Osmotic Breathing
    Psychic Addictions
    Weight Gain -
        Dialectical Weight Problem

BIOLOGISTS AND SCIENTISTS
    Darwin, Charles
    Darwinism
    Gould, Stephen J.
        Mutually-Interpenetrating
        Majesterium→Urban Onion
    Hennig, Willi
    Lamarckian Twist
    Mayer, Ernst

BLOOD
    oxygen conversion rate
        (→Osmotic Breathing)

BODY
    head
        abstract condition of (→
           Temporal Functioning)
    mutation (→'Foyering' Vs. Genetic
        Assumption)
    neck
        giraffe, feeling of (→
           Dimensional
           Plants and Animals)
    physical vs. unphysical sensations
        (→Bodily Nightmares)
    torso
        not arbitrary, for mysterious
           reasons (→Digestion)

BUTTERFLIES
    chance meetings in the wind (→Simultaneity)

CANCER
    Procurement Problem

CHARACTERISTICS
    Gender, Human

CHEMISTRY (see also Appendix)
    brains as extension of (→Tractatus
        of Evolution)
    chemistry of society (→Survival
        Modes of Animals)
    late (relatively) development of
        (→Windfall Physics)
    razor (→Bodily Nightmares)
    Unessential Compounds

CONDITIONS
    Parental Cause

CONSCIOUSNESS
    exceptional rule of (→Decision)
    Thresholds of Consciousness

DARWINISM (See also under Adaptation)
    Amphibians
    Biological Conflict
    Continual Evolution
    Darwin, Charles
    meme-like functionalism (*critique*→
        Introduction)
    Refutation of the Teleological View
        of Evolution
    self-selection (→Biological Substrates)
    spontaneous evolution
            related to consciousness
                    (→Introduction)

DIAGRAMS [SPECIAL DIAGRAMS]
    Animals [Profile]
    Biological Deductions
    Clausal Development [hypothetical]
    Cube of Life [cube]
    Dimensional Plants and Animals (leaf)
    Five New Categories Introduced With Biology
    Fractal Biology [fractals]
    Geometric Variations
            [commonality betw. organs and organics]
    Life Objectives [zig-zag]
    Natural Motions of the Types
            [composite body]
    Parallel Universes [dimensional diagram]
    Plants [Profile]

Quantitative Typology
     [branching of bodies]
Species, Mutual Development of
Survival Modes of Animals [cubes]
Tree of Life [on taxonomy]

DIMENSIONS
     Ascendance & Optimization
     Biological Substrates
     Categorical Genotyping
     Deferential Exception
     Incomplete Information
     Tree of Life
     Windfall Physics

DINOSAURS
     brightly colored (→Introduction)
     crisp vision of (→Decision)
     feathers, not scales (→Introduction)

DISEASE
     cancer (→Procurement Problem)
     Sickness
     Viruses

DISTINCTIONS
     specialization versus perfection
          (→Categorical Dilemma in Biology)

ENERGY
     and reproduction (→Appendix I.
          Animal Phenomenology)
     as dimensional (→Windfall Physics)
     as spiritual theory (→Evasiveness)
     bizarre theories of (→Radical Survivability)
     brain and (→Dimensional Plants and Animals)

compared to consumption and intelligence
(→Biological Deduction)
Compared to mass, complexity, sophistication
(→Reality Effects)
conditional energy (→Alien/ Xenoid Phenotyping)
energy / pleasure interpolator
(→Advanced Brains)
Fractal Biology (and)
Ingestion (and)
latent energy (→Alien/ Xenoid Phenotyping)
Paradox of Prematurity (and)
problem of the phoenix and
(→Exaggerated Birds)
processing (→Immortality)
Volitional Mechanical Guide to Biological Life

EXTRA-TERRESTRIALS:
Alien / Xenoid Gestation
Alien / Xenoid Phenotyping
Brain Centers
Distributed Brains
Extra-Terrestrial Alien Brainstorm
Introduction
Osmotic Breathing
Tractatus of Evolution
Xoration

EVOLUTION: See Darwinism

FUNCTIONS
and meta-functions (→Categorical Genotyping)
argument from profound functions
(→Atomic Perception)
associative (→Categorical Genotyping)
behavioral (→Brain Concept Paradox)

Categorical Adaptation
comparative (→Biological Substrates)
Concealed Functions
correspondence theory of (→Phenomenalism)
corroborative (→Introduction)
dependent function (→Categorical Genotyping)
development of (→Windfall Physics)
exceptional (→Viable Epochranysm)
the four functions (→Animals [Profile])
hierarchical (→Quantitative Typology)
Human function (→Transverse Clausal
                  Development)
ident. appearances used as separate
          (Alien/Xenoid Phenotyping)
info-function (→Advanced Brains)
limit functions (Theories of Incipient Biology)
material and organic (→Psychological
                  Biology)
mental expanding physical (→Psychological
                  Biology)
meta-functions (→Tractatus of Evolution)
mirroring universe (→Tractatus of Evolution)
Paradox of (→Appendix XII.1: Critique of
                  Biology)
Reality Effects
secondary functions in evolution
          (→Tractatus of Evolution)
similar (→Permutation)
social functions, general (→Cube of Life)
social processing (→Barr Body)
Symbiosis
synergistic functions (→Tractatus of Evolution)
unknown (→Introduction)
Viable Epochranysm

GENETICS
  Genetic Deduction
  Genetic Weirdness

GESTATION:
  Alien / Xenoid Gestation

INGESTION
  Negative Ingestion
  Neutral Ingestion
  Positive Ingestion

HOLISM
  Clausal Development
    [Abverse, Transverse, Universal, Nominal]
  Phenomenalism

INSTINCTS
  Animal Intuition
  Atunement
  bizarre theories of (→Reincarnation)
  counter-instinctual (→Evasiveness)
  Evasiveness
  Instincts
  value as a replacement for (→ Contradictory
    Deductions and Elaborate Conclusions)

INTERACTIONS
  Passive -
    Decidualism
  Social -
    Social Argocy

LIFE (See also Monkeys, Butterflies)
       Animals, Profile
       Biological Imperatives
       Five New Categories Introduced with Biology
       Immortality
       Instincts
       Life Objectives
       Meta-Functions of Plants and Animals
       Natural Motions of the Types
       Paradigmatic Animals - Four Types
       Paradox of Prematurity
       Photosynthesis
       Plants, Profile
       Radical Survivability
       Survival Modes of Animals
       Systems Theory Biology
       Theories of Incipient Biology
       Tree of Life
       Volitional Mechanical Guide to Biological Life
       Xoration

MACRO-BIOLOGY
       Clausal Development
          [Abverse, Transverse, Universal, Nominal]
       Entic Property Development Levels

'MANIFESTO'
       Categorical Dilemma in Biology

MONKEYS
       on the islands of Borneo (→Radical Survivability)

Nathan Coppedge

MORPHIZATION
    Categorical Dilemma in Biology

NOMENCLATURE

    Binomial Nomenclature
    Schizoriginal [emerging Age of]
        (→ Clausal Development)
    Trinomial Nomenclature
        (→ Binomial Nomenclature)

OBESITY
    Dialectical Weight Problem

ORGANS
    ALL ANIMALS
        brain
                animal brain as part of
                    human brain (→Animal
                    Intuition)
                Animal Intuition
                Concealed Functions (and)
                lip-brains etc. (→Extra-
                    Terrestrial/ Alien
                    Brainstorm)
            decentralized nervous system
                (→Concealed Functions)
            lungs
                multiple types of
                    (→Osmotic Breathing)
            organs and organization
                (→Geometric Variations)

[ORGANS, CONT'D]

    IN VERTIBRATES
        brain
                ambiguous functions of
                        (→Alien/Xenoid
                        Phenotyping)
                Brain Concept Paradox
                plant brain vs.
                    efficient animals
                        (→Alien / Xenoid
                        Phenotyping)
                primate brain
                        (→Animal Intuition)
        centralized nervous system
                unification of senses
                        (→Bodily Nightmares)
        protostomes & deuterostomes
                        (→Appendix V. Biologist's Gold)

        HUMANS ONLY
        Barr Body
        Brain
                activity and genetics of
                        (→Sensuous Growth)
                and deep traits
                        (→Psychological Biology)
                Gestalt Biology (and)
                mental defenses (→Memetic
                    Exhibition)
                Metaphysics of the Brain
                Nodus - Brain - Synthesis
                quantum (→Higher-Conditional
                    Reality)
                role of (→Higher-Conditional
                    Reality)

soul and (→Nodus - Brain -
Synthesis)
Temporal Functioning
heart
location of (→Permutation)

POST-HUMAN
Advanced Brains
Alien / Xenoid Phenotyping
(some ideas)
Extraterrestrial / Alien Brainstorm
hints about (→Tractatus of Evolution)

PARADIGMS
adaptation (→Introduction)
alien resources (→Fresh Air and Sunlight)
all prior (→Windfall Physics)
animal (→Epoche d' Menage)
AOE (→Biological Conflict)
brains (→Dimensional Plants and Animals)
eating plants (→Biological Eschatology)
electronic (→Ingestion)
environmental degradation (→Dwelling Places)
Four Paradigmatic Animals
Fresh Air and Sunlight
functions (→Higher Functioning)
high and low energy (→Ingestion)
intelligent armies (→Biological Conflict)
life itself (→Entic Property Development Levels)
natural (→Paradigmatic Animals)
need and demand (→Continual Evolution)
organ-related (→Quantum Biology)
paradigmatics (→Introduction,
Adding Concepts to Biology)
realism (→Introduction)
rule-breaking (→Windfall Physics)

success (→Paradigmatic Animals)
survival (→Uber-Adaption)
synthetic (→Epoche d' Menage)
technology (→Introduction)

PARADOXES
Appendix XII: Biological Paradoxes
Brain Concept Paradox
Of survival (→Organic Aberrations)
of upward animals (→Epoché d' Ménàge)
Paradox of Prematurity
Psychological Biology (paradox of)

PERSONALITIES
biological personalities (→Incipient Biology)

PERSPECTIVE
Incomplete Information
Introduction
Profiling of Life Forms

PHENOMENOLOGY (see also Appendix I.)
Phenomenalism

PLANETARY BIOLOGY

criterion of planets (→Modular Biology)

PLANTS
moss (→Universal Flesh)
Plants Profile

PRECONDITIONS
  Antecommunication

PROBLEMS
  addictive problems (→Psychic Addictions)
  as conditioning (→Introduction)
  as definitions (→Introduction)
  biology and philosophy problems
      (→Brain Concept Paradox)
  brain-in-the-vat (→Brain Concept Paradox)
  Categorical Dilemma in Biology
  Fighter-pilot fallacy (→Paradigmatic
Animals
      Four Types)
  Dialectical Weight Problem
  environmental degradation
      (→Dwelling Places)
  Hollow Biology Problem
  Obscure Theses, An Approach
  para-consistency (→Retroactive Retrograde)
  Premium on conscious agents
      (→Nodus, Brain, Synthesis)
  problem of evil (→Darwin, Charles)
  Procurement Problem (cancer studies)
  rationalization (→Reincarnation)
  reductivism (→Atomic Perception)
  solving social problems (→Social Argocy)
  Species Fallacy
  value of consciousness (→Nodus-Brain
-Synthesis)

PSYCHOLOGICAL CONNEXIONS
        Bodily Nightmares
        Psychic Addictions
        Profiling of Life Forms
        Psychological Biology

SEMANTIC BIOLOGY
        Genetic Semantics
        Nodus, Brain, Synthesis

SKIN
        Universal Flesh

SPECIES
        Binomial Nomenclature
        Darwin, Charles
        Species Fallacy

SURVIVAL
        and affective psychology
                (→Bio-Hazards and Other Bro-
mides)
        and automatic understanding
                (→Advertising Skin Deep)
        and biological knowledge (→
Immortality)
        and circumstance
                (→Biological Coincidence)
        and the counter-instinctual (→
Evasiveness)
        and modifications (→Permutation)
        animal phenomenology (Appendix I.)
        bad experiences and (→Appendix XII).
        Biological Deduction

compared to bubbles (→Modular Biology)

compared to God (→Categorical Adaptation)

Dimensional Plants and Animals

dubious advantages (→Categorical Adaptation)

early chances (→Windfall Physics)

enhanced survival (→Radical Survivability)

existential survival (→Introduction)

first principle (→Volitional Mechanical
Guide to Biological Life)

function (→Animals)

general survival categories
(→Categorical Adaptation)

Immortality

necessary vs. statistical (→Volitional Mechanical
Guide to Biological Life)

not celebratory (→*Biologist's Gold*, Appendix
VII.)

Obscure Theses

Origins of Life

principles of (→Modular Biology)

Survival Modes of Animals

'survival species' (→Radical Survivability)

three lemmas of (→Categorical Dilemma
in Biology)

Threshold of (→Paradox of Prematurity)

tough-as-would-have-it
(→Viable Apochrynism)

Uber-Adaption
vs. chemical (→Barr Body)
vs. complexity (→Cube of Life)

SYMBOLS
gossiper (→Brain Concept Paradox)
gun (→Biological Conflict)
Lamarckian Twist
robot (→Biological Conflict)

TAXA [See also Extra-Terrestrials, Functions]
Binomial Nomenclature
Categorical Genotyping
Clausal Development

THEORY (See also under Functions, Theory, Darwinism,
               etc.)
Categorical Adaptation
Categorical Dilemma in Biology
Exceptional Biology
fossil realists (→Butterfly Effect)
fossil formalists (→Butterfly Effect)
Gestalt Biology (see also Introduction)
ideational biology (→Introduction)
Profiling of Life Forms

THEORY CONTRADICTIONS
Categorical Dilemma in Biology
faith in nature (→Categorical Adapta

Nathan Coppedge

tion)

THESES
       Biological Conflict
       Digestion
       identification with environment
              (→Symbiosis)
       Minor susceptibilities (→
              Thresholds of Consciousness)
       minor thesis (→Introduction)
       N-thesis (→Genetics)
       Obscure Theses
       Self-harm advertises material strength
              (→Advertising Skin Deep)
       Systems Theory Biology
       Three theses (→ Introduction)
       virtual biology (*amoebic assimilation*
             →Butterfly Effect)

THOUGHT PATTERNS
       Individual Pluralism

TRIPTYCH MOVEMENTS
       Transmigration, Diaspora, and Settle-
ment

END OF INDEX

Nathan Coppedge

BIO

Nathan Coppedge is the author of over 70 books. He has been quoted in Book Forum and the Hartford Courant, and is a member of the International Honor Society for Philosophy. He is also the author of present and future volumes of the Dimensional Encyclopedia. Volume 3 is Biology.

www.ingramcontent.com/pod-product-compliance
Lightning Source LLC
Chambersburg PA
CBHW051626170526
45167CB00001B/73